Convolutional Neural Networks for Medical Image Processing Applications

Editor

Şaban Öztürk
Assoc. Prof.
Electrical and Electronics Engineering Department
Technology Faculty
Amasya University
Amasya, Turkey

CRC Press is an imprint of the
Taylor & Francis Group, an **informa** business

A SCIENCE PUBLISHERS BOOK

First edition published 2023
by CRC Press
6000 Broken Sound Parkway NW, Suite 300, Boca Raton, FL 33487-2742

and by CRC Press
4 Park Square, Milton Park, Abingdon, Oxon, OX14 4RN

© 2023 Şaban Öztürk

CRC Press is an imprint of Taylor & Francis Group, LLC

Reasonable efforts have been made to publish reliable data and information, but the author and publisher cannot assume responsibility for the validity of all materials or the consequences of their use. The authors and publishers have attempted to trace the copyright holders of all material reproduced in this publication and apologize to copyright holders if permission to publish in this form has not been obtained. If any copyright material has not been acknowledged please write and let us know so we may rectify in any future reprint.

Except as permitted under U.S. Copyright Law, no part of this book may be reprinted, reproduced, transmitted, or utilized in any form by any electronic, mechanical, or other means, now known or hereafter invented, including photocopying, microfilming, and recording, or in any information storage or retrieval system, without written permission from the publishers.

For permission to photocopy or use material electronically from this work, access www.copyright.com or contact the Copyright Clearance Center, Inc. (CCC), 222 Rosewood Drive, Danvers, MA 01923, 978-750-8400. For works that are not available on CCC please contact mpkbookspermissions@tandf.co.uk

Trademark notice: Product or corporate names may be trademarks or registered trademarks and are used only for identification and explanation without intent to infringe.

Library of Congress Cataloging-in-Publication Data (applied for)

ISBN: 978-1-032-10400-3 (hbk)
ISBN: 978-1-032-10401-0 (pbk)
ISBN: 978-1-003-21514-1 (ebk)

DOI: 10.1201/9781003215141

Typeset in Times New Roman
by Radiant Productions

Preface

In addition to facilitating human life, technological developments increase the expectation of the quality of life. For this reason, artificial intelligence supported technological solutions, which are very common in the industry, are rapidly spreading to the medical field today. In this way, it becomes possible to find solutions to many diseases. Today, medical imaging devices have become the most important and widely used method in the diagnosis of diseases. This widespread use brings with it many difficulties in addition to increasing the workload of doctors. In addition to overcoming these difficulties, which are analyzed in detail in this book, early diagnosis, which plays a crucial role in human life, is possible with artificial intelligence methods. For this purpose, this book named 'Convolutional Neural Networks for Medical Image Processing Applications' covers the use of today's most effective artificial intelligence methods for real medical image analysis.

Contents

Preface   iii

1. **Convolutional Neural Networks for Segmentation in Short-Axis Cine Cardiac Magnetic Resonance Imaging: Review and Considerations**   1
   Manuel Pérez-Pelegrí, José V. Monmeneu, María P. López-Lereu and David Moratal

2. **Deep Learning-Based Computer-Aided Diagnosis System for Attention Deficit Hyperactivity Disorder Classification Using Synthetic Data**   34
   Gulay Cicek and Aydın AKAN

3. **Basic Ensembles of Vanilla-Style Deep Learning Models Improve Liver Segmentation from CT Images**   52
   A. Emre Kavur, Ludmila I. Kuncheva and M. Alper Selver

4. **Convolutional Neural Networks for Medical Image Analysis**   75
   Rajesh Gogineni and Ashvini Chaturvedi

5. **Ulcer and Red Lesion Detection in Wireless Capsule Endoscopy Images using CNN**   91
   Said Charfi, Mohamed El Ansari, Ayoub Ellahyani and Ilyas El Jaafari

6. **Do More With Less: Deep Learning in Medical Imaging**   109
   Shivani Rohilla, Mahipal Jadeja and Emmanuel S. Pilli

7. **Automatic Classification of fMRI Signals from Behavioral, Cognitive and Affective Tasks Using Deep Learning**   133
   Cemre Candemir, Osman Tayfun Bişkin, Mustafa Alper Selver and Ali Saffet Gönül

8. **Detection of COVID-19 in Lung CT-Scans using Reconstructed Image Features** 155
   Ankita Sharma and *Preety Singh*

9. **Dental Image Analysis: Where Deep Learning Meets Dentistry** 170
   Bernardo Silva, Laís Pinheiro, Katia Andrade, Patrícia Cury and *Luciano Oliveira*

10. **Malarial Parasite Detection in Blood Smear Microscopic Images: A Review on Deep Learning Approaches** 196
    Kinde Anlay Fante and *Fetulhak Abdurahman*

11. **Automatic Classification of Coronary Stenosis using Convolutional Neural Networks and Simulated Annealing** 227
    Luis Diego Rendon-Aguilar, Ivan Cruz-Aceves, Arturo Alfonso Fernandez-Jaramillo, Ernesto Moya-Albor, Jorge Brieva and *Hiram Ponce*

12. **Deep Learning Approach for Detecting COVID-19 from Chest X-ray Images** 248
    Murali Krishna Puttagunta, S. Ravi and *C. Nelson Kennedy Babu*

Index 267

Chapter 1

Convolutional Neural Networks for Segmentation in Short-Axis Cine Cardiac Magnetic Resonance Imaging
Review and Considerations

Manuel Pérez-Pelegrí,[1] *José V. Monmeneu,*[2] *María P. López-Lereu*[2] and *David Moratal*[1,*]

ABSTRACT

The characterization of the heart function and anatomy requires the segmentation of the main regions at both the systole and diastole. This is usually done by means of magnetic resonance image (MRI) cine sequences, usually with short-axis views of the heart.

Convolutional neural networks (cnn) can be employed to achieve the segmentation of the desired regions of interest. This chapter describes how to design and apply convolutional neural networks for the task of segmentation in short-axis cardiac MRI. It covers the main features that characterize the image modality and an overview of the segmentation problem with (cnn).

---

[1] Center for Biomaterials and Tissue Engineering, Universitat Politècnica de València, Valencia, Spain.
[2] Unidad de Imagen Cardíaca, ERESA-ASCIRES Grupo Biomédico, Valencia, Spain.
* Corresponding author: dmoratal@eln.upv.es

The most popular segmentation models are introduced and the most relevant advances done in them are described. A review of the problem at hand is conducted along an exposition of the current state of the art solutions with cnn in cardiac MRI. Several key elements in the development of cnn for segmentation are discussed, including, selection of loss functions, how to tackle the segmentation problem when small datasets are available, the overfitting problem, approaches for better segmentation depending on the target within the images, and best practices for implementation of segmentation architectures in cine cardiac MRI.

1 Introduction

Heart and cardiovascular diseases are some of the most extended causes of death and morbidity in advanced countries (Townsend et al., 2016). In this context cardiac magnetic resonance imaging (MRI) is an advanced and useful imaging modality (De Roos and Higgins, 2014; Finn et al., 2006) that can give clinicians relevant information of both the structure and the functionality of the heart tissues and is considered the best imaging modality for assessing the heart (La Gerche et al., 2013; Seraphim et al., 2020).

More specifically cardiac short-axis cine MRI acquisitions are especially useful, as they provide a way to visualize different structures of the heart with great contrast and are easy to interpret and to extract segmentation volumes of the regions of interest. However, segmentation is a hard and time consuming task, even with semi-automatic softwares which are often employed in the clinical setting. In this setting convolutional neural networks have shown a great potential to solve the problem of cardiac segmentation in short-axis cine MRI with very accurate results.

Convolutional neural networks (cnn) are a type of deep learning architecture especially designed to treat images. They can be employed to different problem types, and this includes segmentation problems. Convolutional neural networks started to be of interest after the AlexNet convolutional network significantly outperformed all the rival algorithms at the annual ImageNet competition in 2012 (Krizhevsky et al., 2017). Ever since then, convolutional neural networks have increasingly received more attention in the medical imaging field, and in more recent years this also includes cardiac imaging. Segmentation convolutional neural networks have been continuously developed since the first full convolutional neural network (FCN) was described (Long et al., 2015), and nowadays it is one of the fields where deep learning has been more focused in the medical imaging community.

The impact of these algorithms in medical image processing has been growing in later years, and this has also included cardiac MRI. After a search in PubMed (https://pubmed.ncbi.nlm.nih.gov/) with different key words, it can be seen how the growing interest in convolutional neural networks in the medical field has been increasing over the years. Figure 1 shows graphics with the number of publications obtained by year with the key words "convolutional neural networks", "convolutional neural networks segmentation" and "convolutional neural networks cardiac MRI".

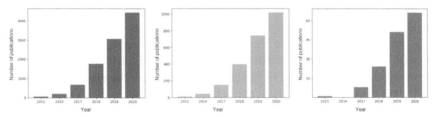

Figure 1: Number of publications by year registered in PubMed with keyword "Convolutional neural networks" (left), "Convolutional neural networks segmentation" (middle) and "Convolutional neural networks cardiac MRI" (right).

In all cases we can see that interest has been growing at an exponential rate for generic, segmentation and specifically for cardiac MRI problems. The exponential growth seems to start clearly around 2017 with no apparent decrease in interest nowadays.

In the following sections of this chapter the problem of segmentation in cardiac short-axis cine MRI with convolutional neural networks is discussed in detail along with important recent advances and applications of interest that could be employed for the design of architectures to segment the regions of interest within the images.

2 Short-Axis Cine MR Imaging

Cine MRI is a type of imaging that is capable of capturing motion. In the case of cardiac MRI, it provides a tool for obtaining images of the heart's motion and enables the visualization of the cardiac cycle. The acquisition of these images is done synchronously with ECG-gating to alleviate the intrinsic movement of the heart. The reconstruction is usually done retrospectively, after a continuous set of simultaneous acquisitions from both the MRI and ECG signals. As the lung also produces motion during the respiratory cycle, the acquisitions are usually done in breath-hold moments. Multiple breath holds are normally required for fully reconstructing the heart in the images.

The most used sequence for cine MRI is SSFP (steady-steady free precession), which is a variation of the gradient-echo sequence. SSFP may be implemented with different configurations and named differently depending on the MRI machine: "balanced FFE" for Philips, "TrueFISP" for Siemens or "FIESTA" for GE as examples. This type of sequence is weighted in T1 and T2, more specifically the signal comes from the ratio between T2 and T1. The images produced have a great contrast between myocardium (which appears dark) and the blood pool (which is bright). The images can be acquired with a great temporal resolution usually in the order of tens of milliseconds. The result is a stack of volumes representing each a frame of the heart cardiac cycle.

In order to characterize the heart motion and function, cine MRIs are usually acquired in the short axis plane (see Fig. 2). These planes allow for the visualization of the left and right ventricle chambers in the same image plane and their contractions are more easily assessed than other planes of the heart anatomy. Figure 3 shows some frames from a slice along the cardiac cycle.

The characterization of the heart motion requires analyzing the tissues at both the end-diastole and end-systole frames (Ishida et al., 2009). In order to do this segmentation a prerequisite is that the main biomarkers are the volumes of the main regions of interest in those frames. The main regions of interest are the left ventricle (LV) cavity (or blood pool), the LV myocardium and the right ventricle (RV) cavity. With these volumes one can assess the mass of the regions and also derive the ejection fraction, which is an important biomarker in assessing the ventricle contractility. Other elements that might be of interest are also visible in these images, mainly the RV myocardium and the papillary muscles. Although the majority of works do not target these regions, and the fact that the blood pool region segmentations tend to include

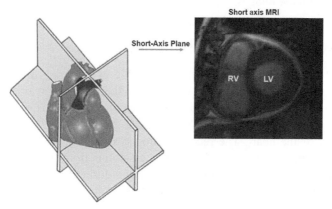

Figure 2: Short axis plane of the heart and a representative view of a slice. RV is the right ventricle and LV the left ventricle. Modified from Cao, 2003.

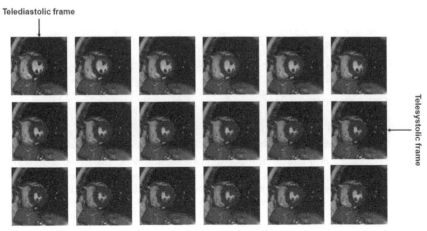

Figure 3: Short-axis slices along the cardiac cycle in a typical acquisition of cine sequences. The images are related in time following the order from left to right, and from up to down. The telediastolic (end of ventricular diastole) and telesystolic (end of ventricular systole) frames are indicated.

the papillary muscles as well, there are some works addressing the importance of the papillary muscles in certain pathologies (Rajiah et al., 2019; Scatteia et al., 2020). Convolutional neural networks have been applied to segment these muscles as well (Bartoli et al., 2021), however there is a considerable lack of work on this specific problem. There could be an increased interest in applying more of these methods to segment the papillary muscles in the near future. Figure 4 presents a short axis in the diastole with all these regions marked and Fig. 5 shows both a typical segmentation of the main regions of interest and a segmentation containing all the mentioned regions.

Besides all the mentioned factors in the cardiac short-axis cine MRI a last consideration is the important differences that exist between the different

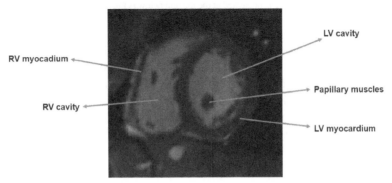

Figure 4: Example of a short axis slice with all the relevant tissue regions within the heart. LV stand for left ventricle and RV right ventricle.

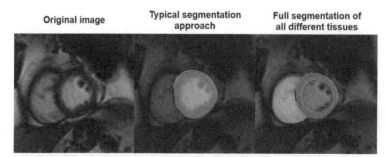

Figure 5: Example of a cardiac short-axis slice and segmentations. The typical segmentation approach to derive the tissue volumes is presented in the central image, this includes the RV cavity, the LV cavity including the papillary muscles and the LV myocardium. The right image presents a segmentation with all major elements segmented, with different label colors for the papillary muscles and the LV inner cavity and an additional label for the RV myocardium.

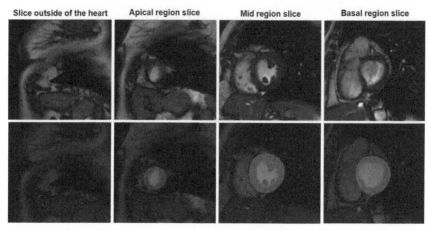

Figure 6: Different regions across a cardiac short-axis volume, including a slice without presence of the heart (left), a slice of the apical region (center-left), a mid region slice (center-right) and a basal slice (right). The original image is presented at the top and below is the typical segmentation approach applied to the main regions.

slices within a volume. The short-axis view can be divided into the apical, mid and basal regions. The apical region is characterized by a low presence of the ventricles within the image, and usually only the LV is present. The Mid region encompasses the central slices of the volume and it clearly presents both the RV and LV in their biggest sizes. Lastly, the basal region represents the upper zone of the volume, with both the RV and LV still present, however there is an increased difficulty in differentiating the RV from other surrounding tissues. Figure 6 shows some slices for all these regions with their respective segmentation.

3 Convolutional Neural Network Architectures for Segmentation

In this section some of the most relevant architectures and layers for the segmentation of neural networks are described. One must consider that this is a brief summary of the most relevant ones, but there exist other architectures and layers that could be employed. In this section some basic convolutional neural network operations are mentioned, which are well described in the literature (Dumoulin and Visin, 2016; Lederer, 2021).

3.1 U-net

The original U-net architecture described by Ronneberger et al., 2015 has become the most popular fully convolutional neural network for medical imaging segmentation. Most of the segmentation problems are tackled with the U-net or a variation of its original design, however, the architecture's core remains in all these variations.

The architecture is similar to that of encoder-decoder architectures (Bank et al., 2020; Vincent et al., 2010). It contains a contracting path where several convolution layers are applied consecutively alongwith pooling operations which reduce the size of the feature maps when the number of feature maps is increased. This path allows the network to find complex patterns at different scales. After reaching the "bottleneck" (the last convolutions of the contracting path) the feature maps go through the up-sampling path which applies up-convolutions consecutively to recover the size lost during the contracting path. Every up-sampled featured stack is additionally concatenated to the output of the contracting path output of the same level, and these are called skip connections. By doing this the network also recovers spatial information that might be lost during the contracting path. After the concatenation more convolutions are applied followed again by the up-convolution, repeating this process until the original input size is reached.

The original U-net was described with the use of convolutions without applying any padding in the borders of the convolution inputs. This meant that in the concatenating section the input from the contracting path had to be cropped to match the spatial size of the up-sampled feature maps. This also meant the final segmentation did not match the image's original size, and represented a centered region of the image. One would want to retain the same size only to apply a padding before every convolution. Figure 7 represents a schematic of the U-net general architecture.

The U-net basic architecture can be easily improved by adding other operations, and most of the implemented U-net architectures have some

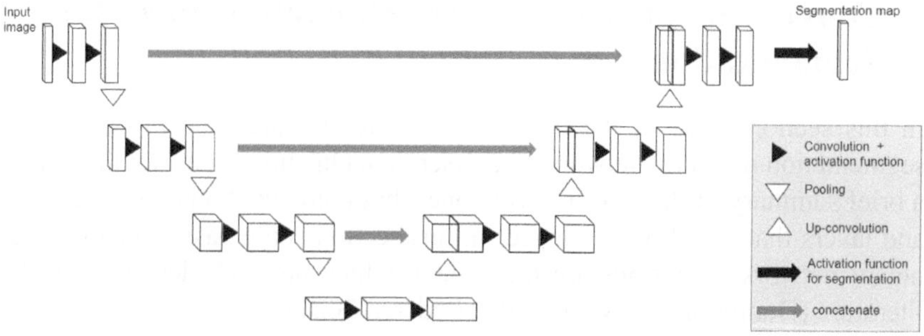

Figure 7: U-net basic architecture scheme. Each block represents the stack of feature maps, which increase in number and reduce in size after each pooling. The process is then reversed through the expanding path with up-convolutions. The concatenation arrows correspond to the skip connections.

degree of modification. The original U-net was intended for 2D images, so it applied 2D convolutions of size 3×3, 2D max-pooling operations and 2D up-convolutions. The activation function was the ReLU (Glorot et al., 2011), which is one of the most commonly applied activation functions in neural networks, and the pooling operation is a maxpooling of 2×2, so the size is halved at every downsampling step. Any of these elements can be substituted by other similar operations, or other operations can be added to suit the problem at hand, for example, it is typical to include a batch normalization operation after every convolution (Ioffe and Szegedy, 2015). Finally, the last layer that generates the segmentation is typically a 1×1 convolution with a sigmoid activation (in cases with only one element to segment) or a softmax activation (in cases with multiple segmentation labels).

3.2 3D U-net

The same idea of the U-net was expanded for 3D image inputs (Çiçek et al., 2016), which are very common in the medical image setting. The most notable difference with respect to the original U-net was the use of 3D convolutions and 3D max-pooling operations. In this case the design described also included batch normalization and the original size was recovered at the end of the up-sampling path.

3.3 V-net

The V-net described by Milletari et al., 2016 included some modifications to the 3D U-net. Each stage included 3 convolutions of size 5×5×5, but the most important changes were the substitution of the pooling operations by strided convolutions and the addition of residual functions.

The strided convolutions used a stride of 2, which ensured that the final feature maps were halved, however employing a strided convolution has the advantage that the downsampling process is also learned, in contraposition to the pooling operation.

The residual operations are a type of function that have demonstrated great potential for improving the performance of neural networks (He et al., 2016a, 2016b). The basic idea is to sum a previous feature map to a convolution output. This operation is especially useful for very deep networks as it can model the identity operation. This can be exemplified with an input x and an operation $h(x)$ that represent the optimal operation to obtain the desired output y. $h(x)$ is a learnable function. The usual convolutional layer represents $h(x)$, so $y = h(x)$. The residual operation adds the input to the convolution output so the final outputs turns out to be, $y = h(x) + x$. When the optimum value for y is equal to x the identity operation is the best option. Usually the initialized weights of the layers have random values close to zero, by adding a sum, in the case that the optimum is close to the identity operation the network will have a much easier way to learn $h(x) = 0$ than to learn $h(x) = x$ and will make the whole training process easier and faster. In the case of the V-net this operation was included at each stage summing the input of the convolutions to the output of the last convolution at that level. Figure 8 has a schematic of how these operations work.

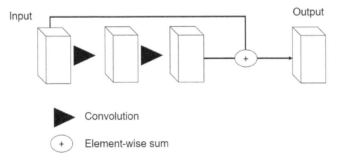

Figure 8: Residual block in convolutional neural networks. The input feature map is summed element-wise to the convolution layers output to obtain the final result.

3.4 Attention U-net

Attention mechanisms were first described for natural language processing in deep learning (Bahdanau et al., 2015). They are operations that allow the network to focus on certain elements within the feature maps. In the case of image processing these have been described in more than one form since the first description of their application to image problems by Xu et al., 2015. A popular attention mechanism applied in medical imaging was described by

Oktay et al., 2018 as an Attention U-net, which allowed the network to learn the segmentation process faster.

The key idea of attention mechanism in convolutional neural networks is to create learnable weight maps that focus on the targeted regions in order to function as pre-segmentations at previous levels and to apply them to certain feature maps. In the work of Oktay et al., 2018 the attention mechanism is defined as a gate that contains certain operations to create a weight map that is multiplied by the output of the contracting path at each level. By doing this the network learns which elements flowing through the skip connections are more useful, and suppresses those less important.

The attention gate contains different operations that use feature maps as inputs that go through the skip connection and the feature map of the previous layer before up-sampling. Figure 9 shows a schematic of all the operations. The most notable parts combine the information from the skip connection input with the feature map of the previous up-sampling level. The result is a feature map with higher weight in the most import spatial regions. This weight map is multiplied channel-wise to the skip connection input to suppress unimportant elements.

An interesting addition to the improvement in performance shown by applying attention gates is the fact that the attention maps generated can be used as a means for explainability on how the neural network works. Explainablity is still a hard problem within the deep learning field as it is still under extensive research (Moreira et al., 2020; Samek et al., 2020). By generating attention maps, it is possible to see which features the neural network gave

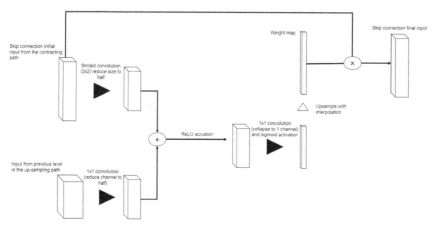

Figure 9: Schematic of a typical attention mechanism implemented in a U-net architecture. Both the feature maps of the contracting path to be passed in the skip connection and the previous feature maps in the expanding map are employed with different operations to generate a weight map that is multiplied (×) by the original input to pass through the skip connection. The whole process is learned by the network to generate a weight map that focuses on the important regions.

more importance to at different scales, helping in the interpretability of how the model generates the segmentations.

3.5 Pyramid Pooling Modules

Pyramid scene parsing modules (PSP), also called pyramid pooling modules have also been described for semantic segmentation problems (Zhao et al., 2017). The idea of these blocks is to reduce the input channel sizes at different scales in parallel to process them independently with convolutional layers, then recovering the size of the reduced feature maps and concatenating the results of each path. This allows for the network to process feature maps at different scales in parallel in contrast to encoder-decoder architectures like the U-net where the information is continuously reduced and then recovered. Figure 10 shows a representation of a typical pyramid pooling module.

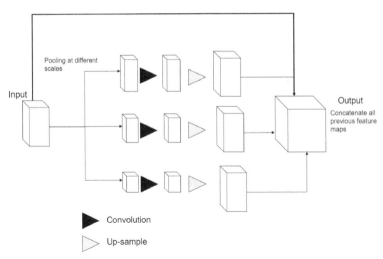

Figure 10: Pyramid pooling module. The input feature maps are downsampled with pooling operations at different scales and processed in parallel with different convolutions. Then the resulting feature maps are up-sampled to the original size and concatenated along the original input to obtain the output feature map.

4 Loss Functions for Segmentation

Segmentation is the problem of classifying each pixel/voxel within an image in a specific category. The loss functions are those used to train the convolution neural networks. They measure the performance of the network while it is being trained by using both the ground truth segmentation and the output offered by the convolutional neural network. In summary, the loss functions give a measurement of the error of the neural network while it is being trained. Through the backpropagation of the loss gradients along each layer (Hecht-Nielsen, 1992) the network improves itself by adjusting its weight in order to minimize the loss value.

This section describes some of the most important loss functions employed for segmentation in convolutional neural networks, their advantages and disadvantages and in which type of situations they should be employed.

4.1 Number of Classes

Before introducing the different loss functions, it should be clarified what binary and multiclass classifications refer to. Binary classification applies to cases where only one region is to be segmented. It is a binary classification because the problem involves working with two classes: the background and the object of interest (or foreground). Usually the target region is labelled as 1 and the background as 0. Multiclass classification refers to the problem of segmentation when different multiple regions within the images need to be segmented. In this case the number of classes would be the number of elements to be segmented plus the background. Segmentation in deep learning assumes that the background represents another class independent of the actual objects desired to be segmented. For multiclass segmentation in cnn the output layers are designed to offer the same number of channels as the number of classes (including the background), where each channel's pixel/voxel has the associated probability for the channel's class. For binary segmentation one channel is enough, as it can encapsulate both the probability associated to the targeted region and to the background, however the output can also be implemented by making the final layer of the network provide a channel for the background and another channel for the target region, although this is rarely done for binary segmentation problems.

4.2 Cross Entropy

The cross-entropy loss is a commonly employed function for classification problems (Murphy, 2012). As such it is also used to tackle segmentation problems. The function considers the ground truth label and the probability associated to the label given by the network. In the case of segmentation, the function returns the average of the associated cross-entropy for each pixel. The equation for binary classification of the cross entropy is represented in Eq. 1 for a single instance classification problem. In this equation y is the target probability (zero or one for classification), and p the calculated probability.

$$CE = -(y\log(p) + (1-y)\log(1-p)) \qquad (1)$$

The cross entropy can be easily extended to multiclass classification, the equation for such cases is presented in Eq. 2, where the final loss is the summation of the cross entropy for all different M labels.

$$CE = -\sum_{c=0}^{M} y_c \log(p_c) \tag{2}$$

When applying the cross-entropy loss for segmentation problems, the final result is the cross entropy averaged for each pixel/voxel within the image. The equation for such cases is presented in Eq. 3 where N is the number of pixels/voxels to segment.

$$CE = \frac{1}{N} \sum_{0}^{N} \sum_{c=0}^{M} y_c \log(p_c) \tag{3}$$

4.3 Dice Coefficient

The Dice coefficient is a measure of the overlap quality between the ground truth segmentation and the predicted segmentation (Zou et al., 2004). The equation of the Dice coefficient is presented in Eq. 4 where A and B represent both the target segmentation map and the segmentation obtained respectively. It is a common score for validating the quality of segmentation. It can also be employed as a loss function, it was first described being used in the V-net (Milletari et al., 2016) and showed that it outperformed other loss functions including the cross-entropy loss. The equation for the Dice loss for binary segmentation problems is presented in Eq. 5, where M is the total number of pixels/voxels to segment, y is the target probability (the real labels), p is the probability associated to the estimated segmentation and μ is a coefficient to avoid dividing by zero. μ is usually selected to have a low value such as 10^{-5} or lower. Extending the Dice loss for multiclass segmentation is trivially accomplished by averaging the Dice score for each label as shown in Eq. 6 where the additional term N represents the number of labels.

$$DSC = 2 \frac{|A \cap B|}{|A| + |B|} \tag{4}$$

$$DL = 1 - 2 \frac{\sum_{C=0}^{M} y_c p_c}{\sum_{C=0}^{M} (y_c + p_c) + \mu} \tag{5}$$

$$DL = 1 - \frac{1}{N} \sum_{0}^{N} 2 \frac{\sum_{C=0}^{M} y_c p_c}{\sum_{C=0}^{M} (y_c + p_c) + \mu} \tag{6}$$

4.4 Weighted Loss Functions

A common problem in the segmentation of medical images is that the sizes of the different labeled regions can be very different, thus resulting in great imbalances that can directly affect the network quality. A big imbalance in the region labels is obtaining good segmentation in the bigger labeled regions while providing bad segmentation in the smaller ones. This can also happen in binary segmentation if the background occupies a much bigger portion of the image.

The way to solve this is to apply a weight factor for each label in the loss function to balance the contribution of all labels. The weighted cross entropy has been utilised by Ronneberger et al., 2015 showing good results. The Dice loss has also been extended in the same way, the weighted Dice loss being named generalized Dice loss which can also improve segmentation results when imbalance is a problem. The Generalized Dice loss was first introduced as a way to better measure the quality of segmentation for unbalanced cases (Crum et al., 2006) and was proposed as a loss function by Sudre et al., 2017.

The weight factor associated with each label is usually calculated taking into account the number of elements that pertain to each label, so the weights are computed using the inverse of the number of pixels/voxels pertaining to the labels and rescaling so the weights add up to one.

The weights could also be manually set for specific labels if one wants to give more importance to a determined labeled region specifically. This may be a better way of balancing in very extreme cases, when there is an extremely small element to segment compared to other regions, as it can create a new imbalance favoring the segmentation of the small object. For binary segmentation this is not problematic as correctly segmenting the foreground will also mean that the background is correct. However, for multiclass cases this can be a real problem.

Consider the hypothetical problem of segmenting the papillary muscles and the myocardium of the LV (a problem which would target the LV muscles) in images with a big field of view. The myocardium occupies a considerably bigger portion of the image compared to the papillary muscles, and the background dominates the image content by a great margin. For the sake of this example assume that the percentage occupation of different image regions, background, myocardium and papillary muscles is 95%, 4.9% and 0.1% respectively. These values are arbitrary but realistic in slices where the papillary muscles are very small. In such a case the loss would be calculated giving weights of 0.001, 0.019 and 0.978 to the background, myocardium and papillary muscles. These weights could make the network fall into a local minimum while training, as the loss is dominated by the papillary muscle segmentation by an extremely big contribution, which could lead to a correct segmentation of the papillary muscles but a bad one for the myocardium. A reasonable approach to this case could be assigning the weight manually. For example, giving a weight of 0.6 to the papillary muscles, 0.35 to the myocardium and 0.05 to the background could alleviate the problem. This new setting will force the network to give a considerably more importance to the papillary muscles due to their small size and higher complexity, but at the same time it gives enough importance to the myocardium for it to have a considerable impact to the loss, which should help the network avoid getting stuck in any undesired local minimum.

As a general rule, as most medical imaging segmentation problems have the problem of imbalance, it is common practice to apply a weighted loss when building convolutional neural networks for segmentation.

4.5 Focal Loss

The focal loss is a variation of the cross entropy and a special type of weighted loss function (Lin et al., 2020). It works by applying an exponent factor and an additional probability term in the equation. Equation 8 presents the new form that the binary cross entropy (Eq. 1) would take, where β is the positive exponent factor. The probability terms are now summarized in p_t (see Eq. 7) as defined by Lin et al., 2020. The exponent factor reduces the loss values of the instances with bigger errors, thus making the network give more importance to the instances where the error is bigger. The key idea of the focal loss can be easily applied to other loss functions apart from the cross-entropy by assigning an exponent factor. A simple extension of this to the Dice loss could be simply to apply the factor to the loss value itself (with factors > 1) which will result in more drastically reduced Dice loss values for worse segmented cases, giving more importance to harder instances.

$$p_t = \begin{cases} p & \text{if } y = 1 \\ 1-p & \text{otherwise} \end{cases} \quad (7)$$

$$FL = -(1-p_t)^\beta \log(p_t) \quad (8)$$

5 Research in Segmentation with Convolutional Neural Networks in Short-Axis Cine MRI

Convolutional neural networks have been researched for the problem of segmentation in short-axis cine MRI in recent years. Some works focus on segmenting a specific region while others target the usual three key regions: LV and RV cavities and LV myocardium. In general, the LV cavity has been shown to be the easiest element to segment, presenting higher quality scores, while the LV myocardium is usually the region with the worst agreement with manual segmentations. The typical approach is to train a neural network for the end-diastole segmentation and another one for the end-systole segmentation, but there are exceptions to this. Also, different works have measured their segmentation performance with different measurements, including derived volume estimations, relative volume errors and segmentation overlap quality measurements. Among the latter most of the works done use the Dice coefficient metric although sometimes other measurements like the Jaccard distance has also been reported. Still, the Dice coefficient remains is the most employed metric to validate the segmentation results in all these works, which is typical in medical image segmentation problems in general.

In the work done by Abdelmaguid et al., 2018, the use of a U-net architecture was tested in different training configurations in order to segment the LV in both systole and diastole in 2D images. They tested different pre-processing steps like intensity normalization in the images, applying an initial region of interest (ROI) and zero-padding. Data augmentation was also tested. They obtained very accurate results with Dice scores around 0.95 for both systole and diastole. Some of their conclusions were that the use of weighted loss functions was necessary to improve the performance, that intensity normalization and applying data augmentation results in better segmentations and that an excessive zero-padding could worsen the network performance.

Another work aimed to segment the LV and RV cavities and the LV endocardium using a novel neural network called Rianet (Tong et al., 2019). It consisted of a typical U-net that included specialized attention blocks. This attention blocks incorporated only information of higher and lower resolution in the contracting path, and the output was then passed into the skip connection as usual. This was a different approach to the classical attention mechanism where the attention map was generated using both information from the contracting and expanding paths. The network consisted of two sub-networks that followed the described structure. The first one detected the region of interest within the image and cropped it. Then, the cropped image was again used as an input to obtain the final segmentation. They reported average Dice scores of 0.94 for the LV cavity, 0.92 for the RV cavity and 0.91 for the LV myocardium.

An interesting approach was taken by Poudel et al., 2017. They implemented a U-net with recurrent layers to segment the LV cavity. Recurrent neural networks are well known for time series analysis and LSTM (Hochreiter and Schmidhuber, 1997) and GRU layers (Cho et al., 2014) have been described as a way to process time-related series. In the proposed work they use a GRU layer at the bottleneck of the U-net. In this case the function of the recurrent layer was not to find temporal relationships, but spatial relationships that were correlated between adjacent slices in the short axis volume, propagating the information throughout all the slices for segmenting the whole stack. They reported the Dice coefficient result of 0.90 and 0.93 for two different datasets.

A novel 2D U-net architecture that incorporated PSP modules (PSPU-net) in the skip connections was described by Perez-Pelegri et al., 2020. The network was tested against a normal 3D U-net for the segmentation of the LV and RV cavities and the LV myocardium. They trained the networks only in end-diastolic frames and obtained Dice scores of 0.95, 0.90 and 0.87 for each respective region, with slightly better results with the PSPU-net. Then they also tested the networks that were trained only in the diastole frames on the end-systolic frames. They obtained worse results, with Dice coefficients around 0.9 and 0.8 for the RV and LV cavities respectively (they did not compare for the myocardium in those cases) and in this case the PSPU-net showed a higher margin of improvement with respect to the 3D U-net. This showed that an improved 2D neural network could perform better than a standard 3D U-net and also that a neural network trained only for the diastole could perform reasonably well with systolic frames. Another important element in this work was that they employed a patch-base analysis, where they selected blocks of 3 slices as inputs of the network instead of entire volumes, with which they could reduce computational costs during the training.

All the previous works were done using specific datasets, however more recent works have also been done in order to apply segmentation with neural networks with different datasets coming from different centers and machine sources in order to solve the problem in a more general setting that could be extended to any image source.

In the work done by Tao et al., 2019 they employed a 2D U-net and trained it in three different settings: with images from the same center and same machine manufacturer with images from different centers from the same manufacturer (multicenter setting) and with images from different centers and different manufacturers (multivendor and multicenter setting). For evaluation they tested each trained U-net on a dataset coming from a different vendor and center than those used for training. In this case they targeted the endocardium and epicardium contours of the LV. They showed that when employing more variable data the final segmentations obtained improved. The network trained with a multivendor and multicenter setting achieved the best results with average Dice coefficients of 0.88, 0.95 and 0.93 for the apical, mid and basal regions of the endocardium respectively, and 0.91, 0.96 and 0.94 for the same region in the epicardium. All the data underwent a preprocessing that included intensity normalization to set all intensity values within the same range, cropping the images so only the center remained, and setting the image resolution to a fixed value of 2×2 mm with a fixed image size of 128×128.

This study showed that including data from different sources could make a network for segmentation more generalizable to other settings.

In another work (Chen et al., 2020) they proposed a data normalization pipeline that included data augmentation to train a 2D U-net with a large dataset coming from a single vendor and center setting and then testing it with other datasets coming from different vendors and centers. The proposed pipeline included resampling the xy plane to a specific resolution of 1.25×1.25 mm without modifying the slice thickness. Intensity normalization was done such that all images had a mean intensity of 0 and a standard deviation of 1 and data augmentation included rotations, flipping operations and zooming effects to artificially increase the heart size. Finally, all images were cropped so the size was constant at 256×256 pixels. The cropping was applied randomly for training but for the test set they standardized the crop only in the central region where the heart is usually present within the images. Employing this pipeline to standardize the training resulted in a network that offered average Dice coefficients around 0.9 for the LV cavity, 0.82 for the LV myocardium and 0.82 for the RV cavity. This showed that standardization on a single center and source machine dataset could result in efficient neural networking for segmentation in datasets coming from different sources, although the quality still falls behind the results obtained when the network is only applied to the same source machine from the same center or when multicenter and multivendor training sets are employed.

A different and novel approach was taken by Pérez-Pelegrí et al., 2021. In this case they trained a U-net neural network to directly predict the LV cavity volume and to segment the region using only the volume values as targets during training. To help the network detect the region to be segmented they added a final scanning module to process the U-net activation map output to force the network to detect circular regions that fitted the target volume. In this case they additionally reduced the resolution of the images to 2×2 mm to alleviate the computations. The neural network was applied to diastolic frames and showed an average relative error of 8% in the predicted volume, while the generated segmentation offered an average Dice coefficient of 0.72. This value improved to 0.79 after some post-processing based on the finding that the resulting segmentation tended to slightly overestimate the mask at the apical regions and to underestimate it at the remaining regions. Although this work's segmentation was not as good as other approaches, it showed that it is possible to obtain reliable segmentations of the LV employing only its volume as a target for a cnn without requiring a previous manual target segmentation, thus offering the possibility to widen the number of potentially useful datasets that lack manual segmentations.

6 Considerations for Cardiac Short-Axis MRI Segmentation with Convolutional Neural Networks

6.1 Dataset Variability

The big advantage of neural networks over other machine learning methods is that their performance scale incredibly well on increasing the training dataset (Ajiboye et al., 2015). Whenever it is possible, increasing the number of samples for training will improve the quality of the segmentation results offered by the network. It is also important that the dataset is a good representation of the real distribution. This means that certain factors play an important role, like the machine source from which the images were acquired. A neural network trained only with images obtained from a single MRI will likely perform worse when applied to images obtained from a different machine. This applies in all senses, meaning that even for the same MRI if the images acquired have very different characteristics, like resolution or field of view, it could impact the neural network training performance. This variability must be taken into account and the amount of images used to train the network must be a good match to the cases it will be applied to after training. If the images have uniform characteristics, then the network will not have problems, but if there are a lot of possible variations, the training dataset should contain enough variability within the images employed for the network to learn these patterns as well.

6.2 Dataset Normalization

Besides the required variability within the dataset, normalization is normally employed to the images as a way to standardize the numerical values within them, their size and their resolution. Not applying normalization will likely make the neural network perform worse and sometimes it may be unavoidable. Applying normalization will also alleviate the problems mentioned when a lot of variability is present in the image dataset.

6.2.1 Resolution Normalization

Normalizing the resolution through the dataset is also recommended in order to improve the network performance. When there are very different resolutions in the images it is usual to resample the image to a specific resolution. The chosen resolution should be good enough for the main spatial features remaining in the image. In the case of short axis cine MRI, a good value for the xy plane would be 1×1 mm or lower. For the slice thickness resolution there can be a great variability between datasets and it can be case dependent (Menchón-Lara et al., 2019), however its resolution is usually lower than in the xy plane. Considering this, it would be reasonable to use the values of the images with the lowest resolution available. Although Isotropic resolution is not necessary, it might be an interesting resampling approximation to work with isotropic voxels, but this will depend on the nature of the problem and the dataset available.

6.2.2 Intensity Normalization

Intensity normalization is an important normalization step. When working with images obtained from different scanners it is crucial as the signal values can differ considerably between images, however it should always be applied even for images coming from the same source machine. Furthermore, normalizing the range of values to low ranges might speed up the training process.

Intensity normalization should always be applied to the whole image volume, even if the network is going to work with single slices. The reason is very simple, if the normalization is applied at the slice level, the intensity of the same tissue within a given volume might vary after slice normalization due to the possible presence of other regions with high contrast present only in certain slices.

There are many possible intensity normalization functions (Reinhold et al., 2019). For datasets coming from the same source a simple min-max normalization is usually enough and has shown to perform well (Abdelmaguid et al., 2018; Perez-Pelegri et al., 2020). Min-max normalization will scale the values of the image so that the minimum is set to zero and the maximum is set to one. For datasets with very different value distributions (like those coming from different machines) the z-score normalization is a better choice. The z-score subtracts each value by the image mean and divides it by the image standard deviation. This will result in a value distribution with zero mean and a standard deviation of one, thus the resulting image value distribution will follow a normal distribution. An additional advantage of z-score normalization over min-max normalization is that it can handle the presence of outlier values within the images better.

6.2.3 Size Normalization

Convolutional neural networks can be designed to work with images of variable size, however when there are pooling operations involved (which is usual in segmentation neural networks) it is necessary that the images have specific sizes to obtain segmentations of the same size as the input image.

For example, for a U-net with n pooling operations that halve the size of the inputs, the image should have a size that can be halved at least n times without having decimals in the division, which means that the image size should be divisible at least by 2^n.

In the case of standardizing the whole dataset to offer images of the same size, cropping and padding operations might be required. Cropping eliminates images borders for size reduction effectively reducing the image field of view. When cropping it is important to ensure that key elements are not erased form the original images. This could happen for example if the heart region appeared near a border. However, as seen in some works a typical approach is to crop the image around the central region to significantly reduce the input size, where the heart is usually placed in the typical short-axis cine acquisitions and is not normally present in the image boundaries.

Padding operations add values to the borders of the image to increase its size. Padding can be applied in different ways: applying a mirror operation to the border region, extending the border with the constant value present at the border or just zero-padding. Padding should be applied with caution, as very big padding factors may reduce the performance of the neural network (Abdelmaguid et al., 2018).

As image size and pixel/voxel resolution go hand to hand, the adequate method to apply both is first to resample the images to the desired resolution and then applying size normalization techniques (Chen et al., 2020). The resolution should be good enough so the images do not lose any meaningful spatial properties and the size should ensure that the field of view is big enough to not lose key elements in the image.

6.3 Overfitting

Overfitting is a common problem in machine learning algorithms (Hawkins, 2004). It is caused when a model adjusts itself too much to the training data while being incapable of generalizing its results to new instances. This translates into a model that has memorized the training data but has not extracted the relevant features that characterize it. In the case of convolutional neural networks this is also a problem, but there are many techniques to avoid it. This section covers some of the most relevant ones.

6.3.1 Use of a Validation Set

A common method to prevent overfitting is to save a small portion of the training dataset to do validation during training. This portion of the dataset is called the validation dataset and is used to evaluate the loss obtained by the network at each epoch during training to see its performance on examples not seen.

Overfitting implies that the network starts to perform too well on the training dataset at the cost of losing generalization capability. Thus a common way to prevent overfitting is to stop training when the validation loss starts to get worse while the training is still improving (Sarle, 1995; Ying, 2019). Such methods are called "early stop methods". An easy approach is to simply stop training once the validation loss increases. However, as the loss can have slight fluctuations across epochs the stopping strategy should not be too strict. It is preferable to define a threshold value for the increase of the validation loss to determine the stopping criteria. The threshold value might depend on the nature of the loss function employed.

6.3.2 Dropout Regularization

Dropout (Srivastava et al., 2014) is a popular regularization technique in neural networks. It works by randomly deactivating neurons in a layer at each training step, the dropout value indicating the proportion of neurons randomly deactivated. This results in a trained network that is akin to an average of trained networks. This property has also been exploited in inferring as the resulting network can work as a type of Bayesian inference (Gal and Ghahramani, 2016).

In the case of convolutional neural networks, dropout is not very common. The main reason is the spatial relationship between the features within each feature map. Randomly setting some features to zero could result in a loss of the spatial relationships. However, there have been some works showing that applying dropout in convolutional neural networks to certain elements can be effectively employed without degrading the performance and gaining regularization properties.

In the work done by Chen et al., 2020, dropout was employed in the skip connections of a U-net architecture. In this case the dropout was applied at a low level of 0.2 and the resulting network worked fairly well even in images from different sources. Other works have shown that applying low dropout values can improve the U-net performance, Colman et al., 2021 showed that a value of 0.2 performed better than not applying dropout, however this was not the case for higher values.

Ghiasi et al., 2018 defined a new method to apply dropout in convolutional neural networks. This is the Dropblock, which applies the dropout to the feature maps in blocks. This means that instead of randomly switching off random features across the entire feature maps, switching off is applied at certain whole regions of the feature map (see Fig. 11). This allows for dropout to be applied more effectively to convolutional neural networks as the blocks

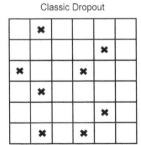 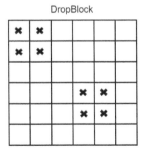

Figure 11: Example of how dropout works on the feature maps in convolutional neural networks. The crossed elements are the features switched off for the next operation. On the left side the typical dropout method is presented, the features are randomly selected across the whole feature map. On the right side the Dropblock method is presented, where the dropout is randomly applied in blocks, allowing for preservation of more spatial relationships between the remaining regions.

affect random regions instead of random features, preserving the spatial relationship in the non-affected parts of the feature map. This method has been applied successfully to the U-net with improved results in medical image segmentation problems (Guo et al., 2019).

As seen in these works, dropout can be employed at certain parts within a convolutional neural network, however it is advisable to consider how the possible loss of spatial relationships can impact the resulting feature maps. Thus, dropout should be applied at very specific sections with very specific methods, like the Dropblock method.

6.3.3 Map Activation Regularization

Another regularization strategy is to apply feature map activity regularization. This consists of penalizing the feature map activation generated by the network. This method not only works as a regularizer, but can also enforce the network to learn more important patterns within the data and suppress the less important ones.

Applying this regularization may require experimentation depending on the dataset and the architecture employed. For example, in the case of a U-net architecture a good option would be to apply the activation penalty at the bottleneck feature maps, as they represent the more abstract and general features found by the network, thus applying the activation penalty at that level forces the network to focus on the more important patterns at the more abstract levels.

Activity regularization can be implemented with a L1 or L2 penalty over the activation maps. L1 will add a factor to the loss function that will penalize the absolute values within the feature map, while L2 will add a factor to the loss penalizing the squared of the values within the feature maps.

Normally this penalty is applied with a weight, as the penalty will be the summed penalty applied to all the values within the feature map, and this can have big impacts on the loss functions and dominate them if a low weight is not applied. There are no specific rules to define the best values in this regard, the regularization setting will depend entirely on the context of the architecture and the problem.

6.4 Data Augmentation

Data augmentation is a common technique applied for training convolutional neural networks. It involves adding new synthetic images to the training dataset and it can serve as a regularization tool as well. The most common way of data augmentation involves applying image modifications on the original training set. Any transformation that preserves the important information of the image can be applied to increase the number of samples in the dataset. Some operations are rotations, translations, zooming, Gaussian noise addition, mirroring or spatial deformations. It is also worthy to mention the possible application of generative convolutional neural networks for generating synthetic data. This includes variational autoencoders (Kingma and Welling, 2014; Pu et al., 2016) and the most successful generative adversarial networks (GANS) defined by Goodfellow et al., 2014. Such networks can be trained to generate new realistic images that can be employed for data augmentation. This has also been successfully applied recently in some works for medical imaging (Frid-Adar et al., 2018).

Applying data augmentation can help the network in many ways. First it can work as a means of regularization and preventing overfitting and it can also make the network insensible to image modifications. For example, image rotations can help the network to find the correct patterns even if the images are rotated, therefore the final model will not be affected by rotated images when inferring segmentations.

As specified, modified images should still preserve information coherence on the image content. This applies differently depending on the problem at hand. In the case of short-axis cardiac cine MRI the majority of transformations could be applied (see Fig. 12 for some examples), however mirroring operations on the horizontal axis should normally not be applied as the image content always presents the RV on the left side, thus flipping such images would create meaningless images where the RV appears on the right side of the image. Such acquisitions are not present in real-life, so applying this transformation is not recommended. In the same sense rotations should not be too big to preserve the relative location of the same tissue. The LV will not be affected by this as it has an approximate circular shape, but the relative

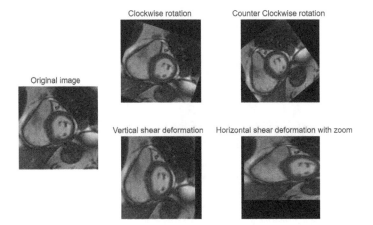

Figure 12: Examples of image modifications to increase the variability of the original image dataset (data augmentation). Any transformation should leave the important regions within the heart to be consistent with its relative location and shape.

position of the RV with respect to it should be of the right side within the image slices if the said region is a target for segmentation.

6.5 Patch-Based Analysis

When aiming to solve a segmentation problem, the best approach is to feed a full image to the cnn and to obtain a full image segmentation. However, sometimes this is not a viable method if the images are very big. Training convolutional neural networks for segmentation can consume a great amount of memory, which is affected by the model size and the input size. Also, the input size will condition the model size. A neural network that uses big images will usually require a bigger number of filters to correctly extract all the features within the image.

A patch-based convolutional neural network takes a small portion of the image as input instead of the full image. Patch-based analysis has been extensively used in segmentation neural networks (Hou et al., 2016; Kao et al., 2020; Ronneberger et al., 2015) in medical imaging problems. This method will reduce the memory requirements and will also speed up the computation, however for inferences the full image segmentation will need to be reconstructed from the different patches passed to the network. Figure 13 shows an example of this for a 2D case. When taking this approach, it is important that the patch size defined can contain enough contextual information, otherwise it might be impossible for the network to find meaningful patterns. In the case of short axis cardiac MRI as the usual target involves either the LV or the RV region (or both at the same time) the patch

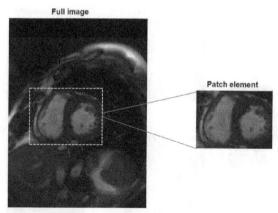

Figure 13: Patch-based analysis for convolutional neural networks. Instead of employing the entire image as input for the network, the image regions are sampled according to a specified size. These patches are then used as inputs for the neural network. The patches should be big enough to encapsulate most of the desired regions to be segmented.

size should be defined to be at least big enough to fit those regions. Leaving a small margin outside their boundaries will also help the networks discriminate them from other non-cardiac regions. Taking this into account the patch size will be entirely determined by the full image size and resolution.

While training with this method one can generate patches in different ways. The patches sampled from an image can be done allowing or disallowing patches to overlap. Also, the sampling could be applied to the entire image or to establish a random number of patches to be extracted from each image. In the latter case, the strategy should also ensure that the target segmented regions are contained in some of the randomly selected patches.

There is another advantage to patch-based analysis besides the memory alleviation. If the regions to be segmented are very small compared to the background this can also help the network detects them. Consider the example of the segmentation of the LV myocardium and papillary muscles exposed in the weighted loss section. In that case, applying a patch-based analysis will reduce the effect of the relative object sizes in the loss, as they will occupy a great portion within the patches.

On the other hand, patch-based analysis also has a disadvantage in cases where the regions to be segmented are small within the image. The problem comes from the imbalance that this can create in the input dataset, where there will be a lot of patches that will not contain any region of interest, and only a small number of patches will contain the targets. Looking at Fig. 13 it can be seen how applying a uniform sampling on the image will generate a large number of inputs that will not contain the heart region, and only one patch will contain the whole portion containing the heart. This creates an imbalance in the training process that can make the network perform badly. In such cases there are other ways to solve the problem: oversampling and undersampling strategies.

6.6 Oversampling and Undersampling

When there is an unbalance within the training dataset samples where a large number of inputs do not contain some of the segmentation targets oversampling and undersampling strategies may be of help for training a segmentation convolutional neural network.

Undersampling involves reducing the number of samples that do not contain the region of interest in order to balance the number of samples containing the segmentation target and the number of samples containing it. It can be useful when the dataset is big and the number of samples containing the region to be segmented is also high. A good and simple method to approach undersampling is to randomly eliminate undesired samples until reaching a number similar to those that contain regions to be segmented.

Oversampling involves increasing the number of samples that contain the region to be segmented to balance their quantity with the samples that do not contain them. There can be different ways to apply oversampling. The simplest one is to create copies of the samples that one wishes to increase. This approach however should be taken with caution as it can potentially lead to overfitting by providing the network with the same input several times during the training. A more reasonable approach is to combine oversampling with data augmentation strategies. By applying some transformations to the oversampled inputs to generate synthetic samples the network will see variable inputs while achieving the desired balance in the training process and will be less prone to suffer from overfitting. In cases with relatively small datasets the latter approach will lead to better results in most cases.

6.7 Batch Normalization

Batch normalization (BN) was first described by Ioffe and Szegedy, 2015. It is a technique that helps normalizing the feature maps in the batch employed for the training process. It is common to apply it after the convolution or after the activation function in neural networks, including convolutional neural networks. Adding BN after each convolution usually makes the network perform better. This technique has also shown to allow for faster training as the number of epochs required is lower due to the network ability to reach an optimum faster. As a general rule it is a good practice to apply BN in convolutional neural networks.

6.8 Batch Size

The batch size refers to the number of samples used in each training step. It is another hyperparameter whose optimum is difficult to estimate before testing different options. Generally, the bigger the batch size, the better is the representation of the dataset at each step of the network, however a low batch size will allow the network to have a larger number of training steps, which could help it to get out of bad local minima and perform better. A relatively big size is often recommended and has shown to be a better option to speed up computation. Furthermore, big batch sizes have been used in practice with good results following specific settings (Hoffer et al., 2017). However, the exact number will depend on the problem at hand and the memory available. In most problems the best option will be a balance between the learning rate and the batch size, as described by Kandel and Castelli, 2020 slightly reducing the learning rate with small or moderate batch sizes (between 16 and 32 input images) producing the best results. These results were specific for a different type of medical imaging problem. Abdelmaguid et al., 2018 tested with different batch sizes for the LV segmentation problem in 2D and concluded that the optimum depended on the selected learning rate. However, they established a global optimum value of 8 images for each batch. It could be considered that this value might be more indicative in the context of cardiac short-axis cine MRI, but it is important to consider that this might depend on the specific settings available.

7 Conclusions

This chapter covered the problem of segmentation in cardiac short-axis cine MRI images using convolutional neural networks. As discussed throughout the text there are many convolutional neural network architectures and layers available for application but in the majority of the literature the most employed architectures consist of U-net and its variations.

The problem of segmentation has been extensively covered with great results in later years but there is still room to improve the quality of the segmentations offered by convolutional neural networks. Some of the most important rules and approaches to consider when building convolutional neural networks for segmentation in cardiac short-axis cine MRI have been described. Many of these approaches are common within the computer vision field and as such they can generally also be considered for other segmentation problems for medical imaging when using convolutional neural networks, however when doing so, specific consideration for the problem will need to be taken as in the problem described herein.

References

Abdelmaguid, E., Huang, J., Kenchareddy, S., Singla, D., Wilke, L. et al. (2018). Left ventricle segmentation and volume estimation on cardiac MRI using deep learning. *ArXiv Computer Vision and Pattern Recognition.* http://arxiv.org/abs/1809.06247.

Ajiboye, A. R., Abdullah-Arshah, R., Qin, H. and Isah-Kebbe, H. (2015). Evaluating the effect of dataset size on predictive model using supervised learning technique. *International Journal of Software Engineering & Computer Sciences (IJSECS),* 1: 75–84. https://doi.org/10.15282/ijsecs.1.2015.6.0006.

Bahdanau, D., Cho, K. H. and Bengio, Y. (2015, September 1). Neural machine translation by jointly learning to align and translate. *3rd International Conference on Learning Representations, ICLR 2015 - Conference Track Proceedings.* https://arxiv.org/abs/1409.0473v7.

Bank, D., Koenigstein, N. and Giryes, R. (2020). Autoencoders. http://arxiv.org/abs/2003.05991.

Bartoli, A., Fournel, J., Bentatou, Z., Habib, G., Lalande, A. et al. (2021). Deep learning–based automated segmentation of left ventricular trabeculations and myocardium on cardiac MR images: A feasibility study. *Radiology: Artificial Intelligence,* 3(1). https://doi.org/10.1148/ryai.2020200021.

Cao, Z. (2003). Simultaneous Reconstruction and 3d Motion Estimation for Gated Myocardial Emission Tomography. University of Florida.

Chen, C., Bai, W., Davies, R. H., Bhuva, A. N., Manisty, C. H. et al. (2020). Improving the generalizability of convolutional neural network-based segmentation on CMR images. *Frontiers in Cardiovascular Medicine,* 7: 105. https://doi.org/10.3389/fcvm.2020.00105.

Cho, K., Van Merriënboer, B., Gulcehre, C., Bahdanau, D., Bougares, F. et al. (2014). Learning phrase representations using RNN encoder-decoder for statistical machine translation. *EMNLP 2014 - 2014 Conference on Empirical Methods in Natural Language Processing, Proceedings of the Conference,* 1724–1734. https://doi.org/10.3115/v1/d14-1179.

Çiçek, Ö., Abdulkadir, A., Lienkamp, S. S., Brox, T., Ronneberger, O. et al. (2016). 3D U-net: Learning dense volumetric segmentation from sparse annotation. *Lecture Notes in Computer Science (Including Subseries Lecture Notes in Artificial Intelligence and Lecture Notes in Bioinformatics),* 9901 LNCS, 424–432. https://doi.org/10.1007/978-3-319-46723-8_49.

Colman, J., Zhang, L., Duan, W. and Ye, X. (2021). DR-Unet104 for multimodal MRI brain tumor segmentation. In *Brainlesion: Glioma, Multiple Sclerosis, Stroke and Traumatic Brain Injuries.*

BrainLes 2020. Lecture Notes in Computer Science (pp. 410–419). Springer, Cham. https://doi.org/10.1007/978-3-030-72087-2_36.

Crum, W. R., Camara, O. and Hill, D. L. G. (2006). Generalized overlap measures for evaluation and validation in medical image analysis. *IEEE Transactions on Medical Imaging*, 25(11): 1451–1461. https://doi.org/10.1109/TMI.2006.880587.

De Roos, A. and Higgins, C. B. (2014). Cardiac radiology: Centenary review. *Radiology*, 273(2), S142–S159. https://doi.org/10.1148/radiol.14140432.

Dumoulin, V. and Visin, F. (2016). A guide to convolution arithmetic for deep learning. http://arxiv.org/abs/1603.07285.

Finn, J. P., Nael, K., Deshpande, V., Ratib, O., Laub, G. et al. (2006). Cardiac MR imaging: State of the technology. In *Radiology* (Vol. 241, Issue 2, pp. 338–354). Radiological Society of North America Inc. https://doi.org/10.1148/radiol.2412041866.

Frid-Adar, M., Diamant, I., Klang, E., Amitai, M., Goldberger, J. et al. (2018). GAN-based synthetic medical image augmentation for increased CNN performance in liver lesion classification. *Neurocomputing*, 321: 321–331. https://doi.org/10.1016/j.neucom.2018.09.013.

Gal, Y. and Ghahramani, Z. (2016). Dropout as a bayesian approximation: Representing model uncertainty in deep learning. *International Conference on Machine Learning*, 1050–1059. http://proceedings.mlr.press/v48/gal16.pdf.

Ghiasi, G., Lin, T. -Y. and Le, Q. V. (2018). DropBlock: A regularization method for convolutional networks. *Advances in Neural Information Processing Systems 31 (NeurIPS 2018), 2018-December*. http://arxiv.org/abs/1810.12890.

Glorot, X., Bordes, A. and Bengio, Y. (2011). Deep sparse rectifier neural networks. *Proceedings of the Fourteenth International Conference on Artificial Intelligence and Statistics*, 315–323. http://proceedings.mlr.press/v15/glorot11a.html.

Goodfellow, I. J., Warde-Farley, D., Lamblin, P., Dumoulin, V., Mirza, M. et al. (2014). Generative adversarial nets. *NIPS'14: Proceedings of the 27th International Conference on Neural Information Processing Systems*, 2672–2680. http://arxiv.org/abs/1308.4214.

Guo, C., Szemenyei, M., Pei, Y., Yi, Y., Zhou, W. et al. (2019). SD-Unet: A structured dropout U-Net for retinal vessel segmentation. *Proceedings—2019 IEEE 19th International Conference on Bioinformatics and Bioengineering, BIBE*, 2019: 439–444. https://doi.org/10.1109/BIBE.2019.00085.

Hawkins, D. M. (2004). The problem of overfitting. *Journal of Chemical Information and Computer Sciences*, 44(1): 1–12. https://doi.org/10.1021/ci0342472.

He, K., Zhang, X., Ren, S. and Sun, J. (2016a). Deep residual learning for image recognition. *Proceedings of the IEEE Conference on Computer Vision and Pattern Recognition (CVPR)*, 770–778. http://image-net.org/challenges/LSVRC/2015/.

He, K., Zhang, X., Ren, S. and Sun, J. (2016b). Identity mappings in deep residual networks. *Lecture Notes in Computer Science (Including Subseries Lecture Notes in Artificial Intelligence and Lecture Notes in Bioinformatics), 9908 LNCS*, 630–645. https://doi.org/10.1007/978-3-319-46493-0_38.

Hecht-Nielsen, R. (1992). Theory of the backpropagation neural network. In *Neural Networks for Perception* (pp. 65–93). Elsevier. https://doi.org/10.1016/b978-0-12-741252-8.50010-8.

Hochreiter, S. and Schmidhuber, J. (1997). Long short-term memory. *Neural Computation*, 9(8): 1735–1780. https://doi.org/10.1162/neco.1997.9.8.1735.

Hoffer, E., Hubara, I. and Soudry, D. (2017). Train longer, generalize better: Closing the generalization gap in large batch training of neural networks. *Advances in Neural Information Processing Systems*, 2017-December: 1732–1742. http://arxiv.org/abs/1705.08741.

Hou, L., Samaras, D., Kurc, T. M., Gao, Y., Davis, J. E. et al. (2016). Patch-based convolutional neural network for whole slide tissue image classification. *Proceedings of the IEEE Computer Society Conference on Computer Vision and Pattern Recognition*, 2016-December: 2424–2433. https://doi.org/10.1109/CVPR.2016.266.

Ioffe, S. and Szegedy, C. (2015). Batch normalization: Accelerating deep network training by reducing internal covariate shift. *32nd International Conference on Machine Learning, ICML 2015*, 1: 448–456. https://arxiv.org/abs/1502.03167v3.

Ishida, M., Kato, S. and Sakuma, H. (2009). Cardiac MRI in ischemic heart disease. *Circulation Journal*, 73(9): 1577–1588. https://doi.org/10.1253/circj.CJ-09-0524.

Kandel, I. and Castelli, M. (2020). The effect of batch size on the generaliability of the convolutional neural networks on a histopathology dataset. *ICT Express*, 6(4): 312–315. https://doi.org/10.1016/j.icte.2020.04.010.

Kao, P. Y., Shailja, F., Jiang, J., Zhang, A., Khan, A. et al. (2020). Improving patch-based convolutional neural networks for MRI brain tumor segmentation by leveraging location information. *Frontiers in Neuroscience*, 13: 1449. https://doi.org/10.3389/fnins.2019.01449.

Kingma, D. P. and Welling, M. (2014, December 20). Auto-encoding variational bayes. *2nd International Conference on Learning Representations, ICLR 2014—Conference Track Proceedings*. https://arxiv.org/abs/1312.6114v10.

Krizhevsky, A., Sutskever, I. and Hinton, G. E. (2017). ImageNet classification with deep convolutional neural networks. *Communications of the ACM*, 60(6): 84–90. https://doi.org/10.1145/3065386.

La Gerche, A., Claessen, G., Van De Bruaene, A., Pattyn, N., Van Cleemput, J. et al. (2013). Cardiac MRI: A new gold standard for ventricular volume quantification during high-intensity exercise. *Circulation: Cardiovascular Imaging*, 6(2): 329–338. https://doi.org/10.1161/CIRCIMAGING.112.980037.

Lederer, J. (2021). Activation functions in artificial neural networks: A systematic overview. http://arxiv.org/abs/2101.09957.

Lin, T. Y., Goyal, P., Girshick, R., He, K., Dollar, P. et al. (2020). Focal loss for dense object detection. *IEEE Transactions on Pattern Analysis and Machine Intelligence*, 42(2): 318–327. https://doi.org/10.1109/TPAMI.2018.2858826.

Long, J., Shelhamer, E. and Darrell, T. (2015). Fully convolutional networks for semantic segmentation. *Proceedings of the IEEE Conference on Computer Vision and Pattern Recognition (CVPR)*, 3431–3440.

Menchón-Lara, R. M., Simmross-Wattenberg, F., Casaseca-de-la-Higuera, P., Martín-Fernández, M., Alberola-López, C. et al. (2019). Reconstruction techniques for cardiac cine MRI. *Insights into Imaging*, 10(1): 1–16. https://doi.org/10.1186/s13244-019-0754-2.

Milletari, F., Navab, N. and Ahmadi, S. A. (2016). V-Net: Fully convolutional neural networks for volumetric medical image segmentation. *Proceedings—2016 4th International Conference on 3D Vision*, 3DV 2016: 565–571. https://doi.org/10.1109/3DV.2016.79.

Moreira, C., Sindhgatta, R., Ouyang, C., Bruza, P., Wichert, A. et al. (2020). An investigation of interpretability techniques for deep learning in predictive process analytics. http://arxiv.org/abs/2002.09192.

Murphy, K. (2012). Machine Learning: A Probabilistic Perspective. MIT.

Oktay, O., Schlemper, J., Folgoc, L. Le, Lee, M., Heinrich, M. et al. (2018, April 11). Attention U-Net: Learning where to look for the pancreas. *Medical Imaging with Deep Learning (MIDL)*. http://arxiv.org/abs/1804.03999.

Perez-Pelegri, M., Monmeneu, J. V., Lopez-Lereu, M. P., Ruiz-Espana, S., Del-Canto, I. et al. (2020). PSPU-Net for automatic short axis cine MRI segmentation of left and right ventricles. *2020 IEEE 20th International Conference on Bioinformatics and Bioengineering (BIBE)*, 1048–1053. https://doi.org/10.1109/bibe50027.2020.00177.

Pérez-Pelegrí, M., Monmeneu, J. V., López-Lereu, M. P., Pérez-Pelegrí, L., Maceira, A. M. et al. (2021). Automatic left ventricle volume calculation with explainability through a deep learning weak-supervision methodology. *Computer Methods and Programs in Biomedicine*, 208: 106275. https://doi.org/10.1016/J.CMPB.2021.106275.

Poudel, R. P. K., Lamata, P. and Montana, G. (2017). Recurrent fully convolutional neural networks for multi-slice MRI cardiac segmentation. *Lecture Notes in Computer Science (Including Subseries Lecture Notes in Artificial Intelligence and Lecture Notes in Bioinformatics)*, 10129 LNCS, 83–94. https://doi.org/10.1007/978-3-319-52280-7_8.

Pu, Y., Gan, Z., Henao, R., Yuan, X., Li, C. et al. (2016). Variational autoencoder for deep learning of images, labels and captions. *Advances in Neural Information Processing Systems*, 2360–2368. http://arxiv.org/abs/1609.08976.

Rajiah, P., Fulton, N. L. and Bolen, M. (2019). Magnetic resonance imaging of the papillary muscles of the left ventricle: Normal anatomy, variants, and abnormalities. *Insights into Imaging*, 10(1): 1–17. https://doi.org/10.1186/s13244-019-0761-3.

Reinhold, J. C., Dewey, B. E., Carass, A. and Prince, J. L. (2019). Evaluating the impact of intensity normalization on MR image synthesis. *Proc. SPIE 10949, Medical Imaging 2019: Image Processing*, 10949: 126. https://doi.org/10.1117/12.2513089.

Ronneberger, O., Fischer, P. and Brox, T. (2015). U-net: Convolutional networks for biomedical image segmentation. *Lecture Notes in Computer Science (Including Subseries Lecture Notes in Artificial Intelligence and Lecture Notes in Bioinformatics)*, 9351: 234–241. https://doi.org/10.1007/978-3-319-24574-4_28.

Samek, W., Montavon, G., Lapuschkin, S., Anders, C. J., Müller, K.-R. et al. (2020). Toward interpretable machine learning: Transparent deep neural networks and beyond. http://arxiv.org/abs/2003.07631.

Sarle, W. (1995). Stopped training and other remedies for overfitting. *Proceedings of the Twenty-Seventh Symposium on the Interface of Computing Science and Statistics*, 352–360.

Scatteia, A., Pascale, C. E., Gallo, P., Pezzullo, S., America, R. et al. (2020). Abnormal papillary muscle signal on cine MRI as a typical feature of mitral valve prolapse. *Scientific Reports*, 10(1): 1–7. https://doi.org/10.1038/s41598-020-65983-1.

Seraphim, A., Knott, K. D., Augusto, J., Bhuva, A. N., Manisty, C. et al. (2020). Quantitative cardiac MRI. *Journal of Magnetic Resonance Imaging*, 51(3): 693–711. https://doi.org/10.1002/jmri.26789.

Srivastava, N., Hinton, G., Krizhevsky, A. and Salakhutdinov, R. (2014). Dropout: A simple way to prevent neural networks from overfitting. *Journal of Machine Learning Research*, 15: 1929–1958.

Sudre, C. H., Li, W., Vercauteren, T., Ourselin, S., Jorge Cardoso, M. et al. (2017). Generalised dice overlap as a deep learning loss function for highly unbalanced segmentations. *Lecture Notes in Computer Science (Including Subseries Lecture Notes in Artificial Intelligence and Lecture Notes in Bioinformatics)*, 10553 LNCS, 240–248. https://doi.org/10.1007/978-3-319-67558-9_28.

Tao, Q., Yan, W., Wang, Y., Paiman, E. H. M., Shamonin, D. P. et al. (2019). Deep learning–based method for fully automatic quantification of left ventricle function from cine MR images: A multivendor, multicenter study. *Radiology*, 290(1): 81–88. https://doi.org/10.1148/radiol.2018180513.

Tong, Q., Li, C., Si, W., Liao, X., Tong, Y. et al. (2019). RIANet: Recurrent interleaved attention network for cardiac MRI segmentation. *Computers in Biology and Medicine*, 109: 290–302. https://doi.org/10.1016/j.compbiomed.2019.04.042.

Townsend, N., Wilson, L., Bhatnagar, P., Wickramasinghe, K., Rayner, M. et al. (2016). Cardiovascular disease in Europe: Epidemiological update 2016. In *European Heart Journal* (Vol. 37, Issue 42, pp. 3232–3245). Oxford University Press. https://doi.org/10.1093/eurheartj/ehw334.

Vincent, P., Larochelle, H., Lajoie, I., Bengio, Y., Manzagol, P. -A. et al. (2010). Stacked denoising autoencoders: Learning useful representations in a deep network with a local denoising criterion. *Journal of Machine Learning Research*, 11: 3371–3408.

Xu, K., Lei Ba, J., Kiros, R., Cho, K., Courville, A. et al. (2015). Show, attend and tell: Neural image caption generation with visual attention. *International Conference on Machine Learning*, 2048–2057. http://proceedings.mlr.press/v37/xuc15.html.

Ying, X. (2019). An overview of overfitting and its solutions. *Journal of Physics: Conference Series*, 1168(2). https://doi.org/10.1088/1742-6596/1168/2/022022.

Zhao, H., Shi, J., Qi, X., Wang, X., Jia, J. et al. (2017). Pyramid scene parsing network. *Proceedings of the IEEE Conference on Computer Vision and Pattern Recognition (CVPR)*, 2881–2890. https://github.com/hszhao/PSPNet.

Zou, K. H., Warfield, S. K., Bharatha, A., Tempany, C. M. C., Kaus, M. R. et al. (2004). Statistical validation of image segmentation quality based on a spatial overlap index. *Academic Radiology*, 11(2): 178–189. https://doi.org/10.1016/S1076-6332(03)00671-8.

Chapter 2

Deep Learning-Based Computer-Aided Diagnosis System for Attention Deficit Hyperactivity Disorder Classification Using Synthetic Data

Gulay Cicek[1,]* and *Aydın AKAN*[2]

ABSTRACT

Attention Deficit Hyperactivity Disorder (ADHD) is a neuropsychiatric disorder that affects children and adults. The fact that ADHD symptoms differ from individual to individual, that similar symptoms are seen in other psychiatric diseases, and that the tests used do not contain objectivity are important ob-

---

[1] Department of Biomedical Engineering, Istanbul University - Cerrahpasa, Department of Software Engineering, Beykent University, Istanbul, TURKEY.
[2] Department of Electrical and Electronics Engineering, Izmir University of Economics, Balcova, Izmir, TURKEY.
* Corresponding author: gulaycicek@beykent.edu.tr

stacles to the correct diagnosis of the disease. It is inevitable to develop robust and reliable tools for the diagnosis of psychiatric diseases such as physical diseases. The role of neuroimaging techniques in the realization of such a robust tool is undeniable. In this study, deep learning-based ADHD classification models were developed with structural MR data. Synthetic data were obtained with online data augmentation techniques. Different data sets were modeled with AlexNet, VggNet, ResNet, SqueezeNet architectures as well as CNN architectures that we developed. The accuracy rate of our architecture, which has a much shorter training period, is over 90%.

1 Introduction

Attention Deficit Hyperactivity Disorder (ADHD) is one of the most common neurodevelopmental disorders affecting children and teenagers (Kahraman and Esra, 2018; Rader et al., 2009). ADHD is characterized by symptoms of inattention, impulsivity, and hyperactivity (Woodard, 2006). The worldwide prevalence of ADHD in children and adolescents is 3.4% (Polanczyk et al., 2015). Just like physical diseases, the diagnosis of psychiatric diseases should be provided with objective and robust tools. The importance of neuroimaging techniques such as Electroencephalogram (EEG), functional magnetic resonance imaging (fMRI) and structural magnetic resonance imaging (sMRI) cannot be denied in the development of such tools. Muhammad et al.'s (Qureshi et al., 2017), anatomical atlas parcellation and cortical parcellation-based features were extracted from sMRI and fMRI images obtained from publicly accessible ADHD-200 datasets. A hierarchical sparse feature elimination algorithm was applied to identify irrelevant features. The accuracy of the ADHD classification model based on SVM is 92.85%. In the study, important brain regions which play an important role in the detection of ADHD, were found based on ANNOVA. Zhu et al., 2017 developed a model for ADHD detection using structural MR. The noise was eliminated with a Gaussian method. The right caudate nucleus, left precuneus and left upper frontal gyrus were extracted with the segmentation method. Classification models were created with three dimensional CNNs. The performance of the model is 62.52%. Zou et al., 2017 developed a deep learning-based ADHD classification model with sMRI and fMRI data obtained from the ADHD-200 datasets. An ADHD classification model was constructed using low-level features. The network was trained by keeping the information in 3rd order tensors. The success rate of the model is 69.15%. Yanli et al. (2019) developed a DEHB classification model with Machine Learning Methods using sMR data obtained from ENIGMA-DEHB dataset. The accuracy of the model is 66.00%. It was determined that intracranial volume, surface area of the brain and subcortical volumes differ between ADHD afflicted and healthy individuals (Zhang-James et al., 2019).

Yuyang et al. (Luo et al., 2020) developed an ADHD classification model using machine learning algorithms using multimodal images. Cortical gray matter thickness and surface area, GM volume of subcortical structures, volume and fractional anisotropy of white matter fiber tracts, and bilateral regional connectivity attributes were classified by giving inputs to community classification algorithms. The accuracy rate of the model created using an objective biomarker is 90%. Kamata et al., 2019 proposed a new classification approach for ADHD auto-diagnosis that uses fractal dimensional complexity mapping on MR data. The classification model was designed with three-dimensional CNN networks and the classification performance of the model is 69.01%. Zhang et al., 2019 presented an ADHD classification model by combining fMRI data with the gcforest method. The functional connectivity and low frequency fluctuation amplitude extracted from fMRI data has been revised by combining it with the gcforest method. The K Nearest Neighbor and combined cascaded forest method were used for the ADHD classification model. Ahmadi et al. (Ahmadi et al., 2021) designed a CNN network automated ADHD system based on EEG signals deep learning approach. A deep convolutional neural network is proposed by extracting spatial and frequency-based features. The highest classification accuracy was achieved with the combination of $\beta 1$, $\beta 2$ and γ bands. The accuracy of the model is 99.46%. In the study, it was determined that the intermediate electrodes made significant contributions to the ADHD classification model.

In studies on ADHD classification, taking steps to overshadow the robustness of the model negatively affected the reliability of the model. For example; unbalanced dataset, not using appropriate validation methods, having few samples in the dataset, not using filters on noise data, and creating models to detect the problem with single feature extraction techniques and classification algorithms are a major obstacle to the model robustness. In this study, we propose a deep learning-based ADHD classification approach with structural MR data. We can list the contributions of our work as follows:

- It was aimed to create an ADHD classification model with a balanced dataset by taking the number of samples of ADHD afflicted and Healthy individuals close to each other.

- The effect of structural MR slices on the ADHD classification model was examined.

- The effects on the ADHD classification model were examined using geometric data augmentation techniques.

- In addition to known CNN architectures such as AlexNet, ResNet, VggNet, the success of the CNN network developed for characterizing structural MR data was examined.
- The accuracy values of the model were calculated with the 5 Fold Cross Validation method.

Other parts of our work are as follows. In the second part of our study, the ADHD diagnosis method is recommended, and the experimental and final results are presented in the third and fourth parts respectively.

2 Method

This study aimed to develop a robust and reliable ADHD detection model with Deep learning methods using structural MR data. For this purpose, a methodology based on deep learning methods for ADHD classification is presented in the study. In this method, the proposed model consists of three stages: dataset acquisition, feature extraction and classification. Two situations were taken into account in the creation of the datasets: (i) for all subjects, all slices of the structural MR data are assigned to a dataset. (ii) slices in which the structural MR data appear completely clear of gray and white matter are determined by a proposed algorithm and assigned to the other dataset. The feature extraction and classification are done automatically according to the CNN architecture used. High classification performance in DL architectures is directly proportional to the amount of data. A large amount of real data may not always be available. Artificial data is produced using data augmentation techniques. In this study, data has been increased by geometric data augmentation techniques. An ADHD classification model was created with AlexNet, VGGNet (16, 19), ResNet (18, 50, 101), and SqueezeNet architectures. Considering the high classification performance we have achieved in DL architectures, our own CNN network has been designed.

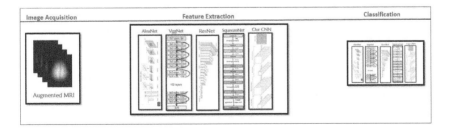

Figure 1: ADHD detection methodology.

2.1 Data Acquisition

Structural MR data was obtained from NPIstanbul NeuroPsychiatry Hospital and ADHD-200 datasets. Detailed information about the data is given in Table 1. Two different datasets were created by assigning all the slices of structural MR to one dataset, and the slices in which gray and white matter are clearly visible to another dataset.

Table 1: NPIstanbul and ADHD200 datasets.

	NPIstanbul Dataset	ADHD-200 Dataset
Healthy	11	157
ADHD	9	178
Total	20	335

The following steps are carried out sequentially in order to characterize the proposed datasets with DL architectures.

- Synthetic data was produced by applying geometric data augmentation techniques.

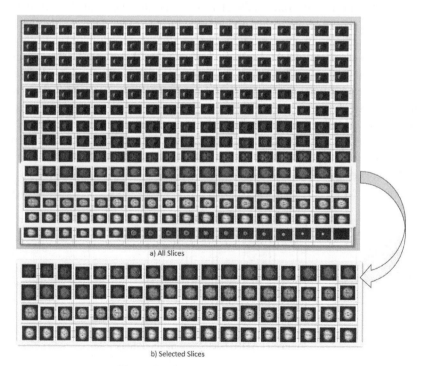

Figure 2: All slices and specific slices.

- Automatic features were extracted and classified using AlexNet, VGGNet (16,19), ResNet (18,50,101) and SqueezeNet architectures.
- 5-fold Cross Validation was performed for random sampling of training and test data.
- Mini-batch accuracy, validation accuracy, mini-batch and validation losses were calculated for the performance measurement of the models.

2.1.1 Data Augmentation

A large number of samples makes an important contribution to robust models. However, sufficient data may not always be available. In this case, the data is augmented by generating synthetic data. There are two types of data augmentation methods: (i) online data augmentation and (ii) offline data augmentation. The online data augmentation method is preferred if sufficient storage is available while offline data augmentation is preferred when there is insufficient storage space and too much artificial data is produced (Andersson and Berglund, 2018). In our study, the effect of geometric data augmentation on ADHD classification models was investigated using CNN networks. For this purpose, data has been increased online with geometric data augmentation techniques such as rotation, scaling and translation. Movement of an object is provided by translation. The object is enlarged with the same scale factor from all directions with scaling. A circular movement of a point connected to the center is provided with rotation. Reflection provides a mirror image of an object. Cutting provides a curved transformation of an object.

2.2 Pre-Processing

Data pre-processing is provided to convert the raw data into a more understandable format by reducing the noise in the data. In this study, the Median filter, which is widely used in digital image processing, was used for noise removal in structural MR slices.

2.3 Deep Learning Approach for Feature Extraction and Classification

Deep Learning (DL) is an artificial neural network inspired by the working mechanism of the human brain (Amin et al., 2021). DL architecture basically consists of input, output, convolution, pooling, relu, and dropout layers (Murphy, 2016). The input layer is responsible for taking the data as input and transmitting it to the convolution layer. The convolution layer is moved over the image to extract the features that will characterize the image. The size of the data is reduced with the pooling layer. Negative values in the data are reduced to zero with the Relu layer. The connections between the nodes are broken, preventing the network from being memorized with the drop out layer. The class of the data is determined by applying the probability function with the softmax layer. The detected class information is transferred to the output layer.

2.3.1 Pre-Trained CNN Models

The AlexNet architecture (Fig. 3), which won the ImageNet 2012 competition, is a kind of convolutional neural network designed by Krizhevsky (Krizhevsky et al., 2012). The network consists of input, convolution, pooling, fully connected and softmax layers. Another type of CNN developed on AlexNet architecture by Simonyan and Zisserman is VGG-16 architecture (Muneeb ul Hassan, 2018). This network detected 14 million images from 10 classes with 92.7% accuracy. The VGG network is deeper than AlexNet due to the increased number of layers. In ResNet architecture, unlike AlexNet and VggNet architectures, instead of increasing the depth, the addition of residual blocks provides a better performance in the network. SqueezeNet is a CNN type that provides a more successful performance than AlexNet with 50 times less parameters.

Figure 3: AlexNet architecture.

2.3.2 Proposed CNN Architecture

The proposed CNN architecture (Fig. 4) is modeled after the 25-layer AlexNet architecture. High classification performance were observed with the AlexNet architecture, which includes all slices of the structural MR. The abbreviated architecture of the AlexNet architecture was designed in order to achieve high classification performance with fewer layers in a shorter time. The aim is to reduce the computational cost and improve classification performance. The MR dimensions taken from different centers are not a fixed size, they are resized with 227*227*3 dimensions and given as input to the proposed network. In the first step, the convolution layer filter size is 5X5 and the step spacing is 3. Following this are the relu and normalization layers. Then, the convolution layer with the first stage features is applied. Then there are the reLU, normalization, fully connected layer, drop out, softmax and classification layer. Our proposed

Figure 4: Proposed CNN architecture.

CNN architecture differs from the AlexNet architecture in that it has a small number of layers (11 layers), a smaller filter size, and a succession of different layers.

2.4 Evaluation Metrics

In this study, the performance of each proposed ADHD detection model was evaluated in terms of training accuracy, validation accuracy, training loss, and validation loss. While the training accuracy refers to the ability of the model to correctly distinguish the classes of these samples by applying the examples used in the training process to the model created for the determination of the problem, the training loss expresses the opposite situation, namely the inability to distinguish the classes correctly. While validation accuracy refers to the ability of the model to correctly distinguish the classes of these samples by applying the examples that are not used in the training process to the model created for the determination of the problem, the validation loss expresses the opposite situation, namely the inability to distinguish the classes correctly.

In this study, 5 Fold Cross Validation is used. It is a widely used procedure to evaluate and compare the performance of machine learning algorithms. In cross validation, a hyper parameter "k" determines the number of subsets of the dataset. With the determination of the K hyper parameter, the training and test sets are determined. The working steps of the method are as follows.

- The dataset is divided into k subsets.
- The first subset is used for testing, the other, $k - 1$ subsets for training.
- The accuracy of the model is and the error value are calculated back.
- The second and third steps, are performed by changing the test subset in each respectively.
- Average accuracy of all k trials is calculated.

3 Experimental Results

ADHD classification experiments were performed on MATLAB® 2019 8GB RAM, Intel i5-8265U CPU. The proposed slices selected with the algorithm were trained on CNN networks. The SGD optimizer with a learning rate of 0.001 was used. Since the input dimensions of the images differ according to the architecture used, the data was resized according to the architecture used. For example, while the input sizes in AlexNet architecture are 227*227*3, the input sizes in SqueezeNet architecture are determined as 224*224*3. After several trials, during the training phase, the hyperparameters were determined as follows. Initial Learning rate was selected as 0.001, Momentum 0.9, L2Regularization 1e-4, GradientThresholdMethod L2norm, MaxEpochs 40, MiniBatchSize 128, ValidationFrequency 10 and ValidationPatience 8.

Poor performance results were observed with the ResNet18 (Table 3), ResNet50 (Table 4), ResNet101 (Table 2.4), VGG-16 (Table 6), VGG-19

Table 2: Performance results of augmented data with AlexNet architecture.

Fold	Epoch	Iteration	DataSet-1 [Specific Slices] T Acc.	V Acc.	T Loss.	V Loss.	DataSet-2 [All Slices] T Acc.	V Acc.	T Loss.	V Loss.
Fold1	1	1	52.11	74.65	1.32	0.66	45.07	57.75	1.27	0.82
	10	10	74.65	80.28	0.50	0.42	73.24	83.10	0.62	0.38
	20	20	74.65	80.28	0.50	0.42	**85.92**	**92.96**	**0.27**	**0.19**
	30	30	78.87	78.87	0.42	0.41	**94.37**	**100.0**	**0.12**	**0.09**
	40	40	80.28	80.28	0.40	0.38	**97.18**	**100.00**	**0.07**	**0.05**
Fold2	1	1	59.15	67.61	1.05	0.60	56.34	47.89	1.52	1.06
	10	10	77.46	78.87	0.63	0.42	77.46	84.51	0.52	0.38
	20	20	76.06	80.28	0.42	0.40	**88.73**	**95.77**	**0.26**	**0.17**
	30	30	80.28	81.69	0.38	0.38	**97.18**	**98.59**	**0.11**	**0.06**
	40	40	78.87	81.69	0.39	0.37	**98.59**	**98.59**	**0.03**	**0.0**
Fold3	1	1	47.89	67.61	1.22	1.52	52.11	47.89	1.98	1.98
	10	10	66.20	57.75	0.58	0.55	73.24	80.28	0.62	0.46
	20	20	78.87	77.46	0.47	0.43	**87.32**	**91.55**	**0.30**	**0.28**
	30	30	81.69	78.87	0.35	0.39	**95.77**	**98.59**	**0.15**	**0.09**
	40	40	78.87	80.28	0.42	0.38	**98.59**	**100.0**	**0.06**	**0.03**
Fold4	1	1	59.15	67.61	1.10	0.67	57.75	56.34	1.04	0.64
	10	10	71.83	71.83	0.56	0.54	59.15	70.42	0.67	0.51
	20	20	74.65	78.87	0.46	0.41	**94.37**	**97.18**	**0.16**	**0.13**
	30	30	69.01	78.87	0.52	0.39	**94.37**	**98.59**	**0.11**	**0.03**
	40	40	83.10	78.87	0.39	0.39	**100.0**	**100.0**	**0.01**	**0.09**
Fold5	1	1	49.30	38.03	2.09	0.80	52.11	53.52	1.17	1.43
	10	10	73.24	74.65	0.79	0.51	64.79	81.69	0.63	0.46
	20	20	77.46	77.46	0.43	0.42	**85.92**	**95.77**	**0.36**	**0.24**
	30	30	77.46	81.69	0.42	0.39	**98.59**	**100.0**	**0.12**	**0.06**
	40	40	80.28	81.69	0.40	0.36	**100.0**	**100.0**	**0.03**	**0.01**

Table 3: Performance results of augmented data with ResNet-18 architecture.

Fold	Epoch	DataSet-1 [Specific Slices]				DataSet-2 [All Slices]			
		T Acc.	V Acc.	T Loss.	V Loss.	T Acc.	V Acc.	T Loss.	V Loss.
Fold1	1	51.56	52.11	1.36	0.76	41.41	57.75	1.25	0.91
	10	78.91	74.65	0.41	0.55	46.88	54.93	0.85	0.81
	20	82.81	73.24	0.39	0.49	56.25	61.97	0.71	0.76
	30	85.94	73.24	0.29	0.50	60.94	61.97	0.65	0.69
	40	86.72	77.46	0.34	0.49	50.78	59.15	0.81	0.70
Fold2	1	56.25	54.93	0.94	0.66	51.56	53.52	1.02	0.84
	10	76.56	78.87	0.53	0.52	56.25	52.11	0.69	0.75
	20	82.03	76.06	0.40	0.49	50.00	54.93	0.79	0.76
	30	85.94	77.46	0.32	0.50	51.56	54.93	0.80	0.82
	40	82.03	71.83	0.39	0.71	53.13	54.93	0.80	0.88
Fold3	1	38.28	32.39	1.61	0.89	50.00	47.89	1.24	0.88
	10	71.09	76.06	0.46	0.56	53.13	47.89	0.75	0.85
	20	85.94	73.24	0.36	0.46	58.59	53.52	0.72	0.73
	30	85.16	77.46	0.34	0.45	55.47	50.70	0.83	0.78
	40	78.91	73.24	0.39	0.44	50.00	54.93	0.96	0.75
Fold4	1	43.75	46.48	1.84	1.17	39.84	49.30	1.12	0.84
	10	74.22	53.52	0.46	0.68	64.84	53.52	0.73	0.81
	20	78.91	70.42	0.42	0.55	53.91	54.93	0.74	0.70
	30	80.47	71.83	0.41	0.66	50.78	54.93	0.91	0.73
	40	85.16	70.42	0.36	0.56	61.72	57.75	0.68	0.76
Fold5	1	49.22	64.79	1.10	0.72	55.47	42.25	0.84	0.88
	10	73.44	66.20	0.48	0.57	54.69	50.70	0.73	0.76
	20	80.47	73.24	0.40	0.50	47.66	56.34	0.89	0.73
	30	86.72	71.83	0.31	0.58	66.41	56.34	0.60	0.71
	40	84.38	73.24	0.39	0.50	64.06	59.15	0.64	0.68

(Table 7), SqueezeNet (Table 8) architectures used to characterize the 1st and 2nd datasets. Over 90.00% accuracy was observed with the AlexNet architecture (Table 2), which was used to characterize the 2nd dataset containing all slices. The high classification success achieved with this architecture has led us to benefit from it while designing our own CNN network (Table 9). Although our proposed CNN network is similar to the AlexNet architecture, it differs from it with features such as fewer layers, smaller filter sizes, and a succession of different layers. The low number of layers did not cause any reduction in performance, on the contrary, it exhibited higher classification success, reduction in training time, and fast operations. Due to its high classification ability and the advantageous features it brings, it is hoped that it will be the preferred CNN network for ADHD detection.

Table 4: Performance results of augmented data with ResNet-50 architecture.

Fold	Epoch	\multicolumn{4}{c}{DataSet-1 [Specific Slices]}	\multicolumn{4}{c}{DataSet-2 [All Slices]}
		T Acc.	V Acc.	T Loss.	V Loss.	T Acc.	V Acc.	T Loss.	V Loss.
Fold1	1	36.72	46.48	0.81	0.69	44.53	52.11	0.93	0.72
	10	76.56	78.87	0.49	0.51	46.09	54.93	0.90	0.69
	20	82.03	52.11	0.41	0.60	61.72	59.15	0.67	0.19
	30	80.47	61.97	0.40	0.54	58.59	59.15	0.74	0.69
	40	78.13	80.28	0.42	0.48	60.16	63.38	0.72	0.63
Fold2	1	57.03	57.75	0.73	0.66	53.91	38.03	0.79	0.88
	10	67.97	73.24	0.65	0.50	60.16	46.48	0.67	0.76
	20	81.25	67.61	0.40	0.54	57.81	53.52	0.71	0.77
	30	83.59	70.42	0.40	0.61	64.06	57.75	0.68	0.74
	40	83.59	74.65	0.34	0.53	61.72	53.52	0.66	0.79
Fold3	1	50.78	70.42	0.96	0.65	45.31	45.25	0.82	0.78
	10	70.31	71.83	0.53	0.52	57.03	54.93	0.76	0.74
	20	78.13	70.42	0.45	0.48	49.22	53.52	0.82	0.80
	30	78.13	78.87	0.43	0.39	57.03	60.56	0.69	0.69
	40	82.81	81.69	0.36	0.37	59.38	64.79	0.71	0.65
Fold4	1	51.56	66.20	0.75	0.67	54.69	57.75	0.75	0.73
	10	78.13	70.42	0.46	0.55	57.03	53.52	0.71	0.75
	20	79.69	76.06	0.44	0.49	59.38	60.56	0.70	0.70
	30	74.22	74.65	0.43	0.47	55.47	57.75	0.73	0.66
	40	83.59	77.46	0.36	0.46	66.41	56.64	0.59	0.63
Fold5	1	67.19	67.61	0.58	0.69	52.34	38.03	0.88	0.83
	10	80.47	60.56	0.50	0.66	52.34	60.56	0.77	0.70
	20	75.00	61.97	0.53	0.66	54.69	54.93	0.71	0.72
	30	83.59	69.01	0.32	0.62	64.06	66.20	0.66	0.68
	40	78.91	56.34	0.40	0.64	57.81	61.97	0.67	0.74

In conclusion, we can summarize the results obtained from the experiments as follows.

- When all slices are trained with AlexNet architecture, the performance increases as the number of iterations increases (highest accuracy is 100.00%), when selected sections are trained, the performance increases as the number of iterations increases (highest accuracy is 80.28%), however, performance results comparable with all sections are not observed.

- When all slices are trained with ResNet architecture, the performance increases with the number of iterations (the highest accuracy rate is 64.00%), however, performance results as high as selected sections (highest accuracy is 86.72%) are not observed.

Deep Learning Based-Computer-Aided Diagnosis System for ADHD ■ 45

Table 5: Performance results of augmented data with ResNet-101 architecture.

		DataSet-1 [Specific Slices]				DataSet-2 [All Slices]			
Fold	Epoch	T Acc.	V Acc.	T Loss.	V Loss.	T Acc.	V Acc.	T Loss.	V Loss.
Fold1	1	48.44	71.83	1.03	0.57	41.41	57.75	1.25	0.91
	10	75.78	81.69	0.50	0.44	46.88	54.93	0.85	0.81
	20	82.81	80.28	0.38	0.37	56.25	61.97	0.71	0.76
	30	82.81	81.69	0.35	0.35	60.94	61.97	0.71	0.76
	40	79.69	77.46	0.39	0.36	50.78	59.15	0.81	0.70
Fold2	1	49.22	50.70	0.96	0.72	51.56	53.52	1.02	0.84
	10	56.25	52.11	0.69	0.85	56.25	52.11	0.69	0.85
	20	79.69	70.42	0.44	0.57	50.00	54.93	0.79	0.76
	30	80.47	66.20	0.40	0.54	51.56	54.93	0.80	0.82
	40	87.50	54.93	0.29	0.55	53.13	54.93	0.80	0.88
Fold3	1	46.88	52.11	0.74	0.69	50.00	47.89	1.24	0.88
	10	71.88	81.69	0.49	0.42	53.13	47.89	0.75	0.85
	20	79.69	57.75	0.42	0.59	58.59	53.52	0.72	0.73
	30	76.56	61.97	0.41	0.58	55.47	50.70	0.83	0.78
	40	82.03	84.51	0.38	0.41	50.00	54.93	0.96	0.75
Fold4	1	47.66	66.20	0.73	0.67	39.84	49.30	1.12	0.84
	10	83.59	76.06	0.39	0.55	64.84	53.52	0.73	0.81
	20	78.91	77.46	0.50	0.52	53.91	54.93	0.74	0.70
	30	82.03	76.06	0.36	0.50	50.78	54.93	0.91	0.73
	40	84.38	78.87	0.34	0.44	61.72	57.75	0.68	0.76
Fold5	1	76.56	80.28	0.50	0.37	55.47	42.25	0.84	0.88
	10	72.66	52.11	0.53	0.67	54.69	50.70	0.73	0.76
	20	79.69	81.69	0.42	0.46	47.66	56.34	0.89	0.73
	30	82.03	77.46	0.38	0.47	66.41	56.34	0.60	0.71
	40	76.56	80.28	0.50	0.37	64.06	59.15	0.64	0.68

- When all slices are trained with ResNet-50 architecture, the performance increases with the number of iterations (highest accuracy rate 66.20%). however, performance results as high as selected sections (highest accuracy rate 83.59%) are not observed.

- When all slices are trained with Vgg-16 architecture, the performance increases steadily as the number of iterations increases (highest accuracy is 59.38%) and lower performance results (highest accuracy is 51.56%) are observed with training of selected sections.

- When all slices are trained with VGG-19 architecture, the performance increases steadily as iterations increase (highest accuracy 51.56%) and lower high performance results (highest accuracy 51.56%) are observed in selected sections.

Table 6: Performance results of augmented data with VGG-16 architecture.

Fold	Epoch	DataSet-1 [Specific Slices]				DataSet-2 [All Slices]			
		T Acc.	V Acc.	T Loss.	V Loss.	T Acc.	V Acc.	T Loss.	V Loss.
Fold1	1	56.25	52.11	6.47	7.63	46.09	52.11	8.23	7.63
	10	44.53	47.89	0.69	0.69	58.59	52.11	0.69	0.69
	20	49.22	47.89	0.69	0.69	45.31	52.11	0.69	0.69
	30	44.53	47.89	0.69	0.69	52.34	52.11	0.69	0.69
	40	43.75	47.89	0.69	0.69	48.44	52.11	0.69	0.69
Fold2	1	46.88	52.11	8.42	7.63	53.13	47.89	7.45	8.30
	10	46.88	47.89	0.69	0.69	53.91	47.89	6.38	8.30
	20	45.31	47.89	0.69	0.69	39.06	47.89	8.98	8.30
	30	50.00	47.89	0.69	0.69	38.28	52.11	8.60	7.63
	40	39.06	47.89	0.69	0.69	52.34	52.11	6.70	2.95
Fold3	1	48.44	47.89	8.10	8.30	39.84	52.11	9.45	7.63
	10	41.41	47.89	0.69	0.69	50.78	47.89	0.69	0.69
	20	50.78	47.89	0.69	0.69	48.22	53.52	0.82	0.80
	30	46.09	47.89	0.69	0.69	57.03	60.56	0.69	0.69
	40	43.75	47.89	0.69	0.69	59.38	64.79	0.71	0.65
Fold4	1	50.00	53.52	7.84	7.40	55.47	46.48	6.99	8.53
	10	47.66	46.48	0.69	0.69	53.91	53.52	0.69	0.69
	20	47.66	46.48	0.69	0.69	50.78	53.52	0.69	0.69
	30	48.44	46.48	0.69	0.69	50.00	53.52	0.69	0.69
	40	47.66	46.48	0.69	0.69	49.22	53.52	0.69	0.69
Fold5	1	50.78	49.30	4.64	8.08	44.53	53.52	8.82	7.40
	10	52.34	46.48	0.69	0.69	52.34	60.56	0.77	0.70
	20	51.56	46.48	0.69	0.69	54.69	54.93	0.71	0.72
	30	46.88	46.48	0.69	0.69	64.06	66.20	0.66	0.68
	40	51.56	46.48	0.69	0.69	57.81	61.97	0.67	0.74

- A steady increase in performance (highest accuracy 53.13%) is observed when all slices are trained with SqueezeNet arhitecture and selected slices (highest accuracy 50.0%) have lower results compared to all slices.

- When all slices are trained with the CNN architecture we designed, the performance (highest accuracy 100.0%) increases steadily with the number of iterations, but such high performance results in selected sections (highest accuracy 71.83%) are not observed.

As a result, higher performance results were obtained with the CNN network we designed. This success has been observed with the use of whole sections.

Table 7: Performance results of augmented data with VGG-19 architecture.

		DataSet-1 [Specific Slices]				DataSet-2 [All Slices]			
Fold	Epoch	T Acc.	V Acc.	T Loss.	V Loss.	T Acc.	V Acc.	T Loss.	V Loss.
Fold1	1	32.03	52.11	10.44	7.63	55.47	52.11	7.03	0.63
	10	46.09	47.89	0.69	0.69	55.47	52.11	7.03	7.63
	20	50.00	47.89	0.69	0.69	58.59	52.11	0.68	0.69
	30	47.66	47.89	0.69	0.69	58.59	52.11	0.68	0.69
	40	48.44	47.89	0.69	0.69	45.31	52.11	0.69	0.69
Fold2	1	49.22	52.11	7.41	7.63	51.56	47.89	7.64	8.30
	10	48.44	47.89	0.69	0.69	60.94	52.11	7.34	7.63
	20	48.44	47.89	0.69	0.69	42.19	47.89	8.71	8.30
	30	46.88	47.89	0.69	0.69	55.47	52.11	6.75	8.30
	40	48.44	47.89	0.69	0.69	41.41	52.11	8.66	7.63
Fold3	1	35.16	52.11	10.19	7.63	43.75	52.11	8.76	7.63
	10	43.75	47.89	0.69	0.69	57.03	54.93	0.76	0.74
	20	49.22	47.89	0.69	0.69	49.22	53.52	0.82	0.80
	30	47.66	47.89	0.69	0.69	57.03	60.56	0.69	0.69
	40	47.66	47.89	0.69	0.69	59.38	64.79	0.71	0.65
Fold4	1	52.34	46.48	7.49	8.53	50.00	53.52	7.82	7.40
	10	50.00	46.48	0.69	0.69	46.09	46.48	8.59	8.53
	20	45.31	46.48	0.69	0.69	52.34	53.52	7.59	7.40
	30	45.31	46.48	0.69	0.69	50.78	53.52	7.84	7.40
	40	46.09	46.48	0.69	0.69	40.63	46.48	9.19	4.76
Fold5	1	27.34	53.52	11.38	7.40	50.78	53.52	7.64	7.40
	10	48.44	46.48	0.69	0.69	52.34	60.56	0.77	0.70
	20	43.75	46.48	0.69	0.69	48.44	53.52	8.22	7.40
	30	46.88	46.48	0.69	0.69	53.13	53.52	7.17	7.40
	40	51.56	46.48	0.69	0.69	50.78	53.52	7.35	7.40

4 Conclusion

ADHD is a neuropsychiatric disorder. As with all diseases, early diagnosis is very important for ADHD. In such psychiatric diseases, delay in diagnosis causes many psychiatric diseases accompanying it, increasing the difficulty of diagnosis and negatively affecting every aspect of the individual's life. Neuroimaging techniques play an important role in the diagnosis of psychiatric diseases as well as physical diseases. Significant changes were observed in brain sMRI of individuals with ADHD when compared to healthy individuals. Volumetric reduction was observed in cortical thickness, gray and white matter, white matter, prefrontal cortex, and basal ganglia sections.

In this study, a deep learning-based ADHD classification model was developed with structural MR data. Two separate datasets were created by assign-

Table 8: Performance results of augmented data with SqueezeNet architecture.

		\multicolumn{4}{c}{DataSet-1 [Specific Slices]}	\multicolumn{4}{c}{DataSet-2 [All Slices]}
Fold	Epoch	T Acc.	V Acc.	T Loss.	V Loss.	T Acc.	V Acc.	T Loss.	V Loss.
Fold1	1	43.75	47.89	8.88	8.30	46.88	52.11	8.44	7.63
	10	50.00	47.89	0.69	0.69	42.97	47.89	0.69	0.69
	20	42.97	47.89	0.69	0.69	46.88	47.89	0.69	0.69
	30	41.41	47.89	0.69	0.69	46.88	47.89	0.69	0.69
	40	50.00	47.89	0.69	0.69	50.00	47.89	0.69	0.69
Fold2	1	57.03	52.11	3.05	7.63	42.19	52.11	9.21	7.63
	10	50.00	47.89	0.69	0.69	45.31	47.89	0.69	0.69
	20	46.88	47.89	0.69	0.69	50.00	47.89	0.69	0.69
	30	42.97	47.89	0.69	0.69	47.66	47.89	0.69	0.69
	40	50.00	47.89	0.69	0.69	49.22	47.89	0.69	0.69
Fold3	1	47.66	47.89	7.96	8.30	40.63	52.11	9.35	7.63
	10	47.66	47.89	0.69	0.69	47.66	47.89	0.69	0.69
	20	48.44	47.89	0.69	0.69	51.56	47.89	0.69	0.69
	30	45.31	47.89	0.69	0.69	47.66	47.89	0.69	0.69
	40	45.31	47.89	0.69	0.69	49.22	47.89	0.69	0.69
Fold4	1	48.44	53.52	5.74	7.40	50.00	46.48	7.84	8.53
	10	45.31	46.48	0.69	0.69	49.22	46.48	0.69	0.69
	20	50.78	46.48	0.69	0.69	51.56	46.48	0.69	0.69
	30	42.97	46.48	0.69	0.69	50.00	46.48	0.69	0.69
	40	42.97	46.48	0.69	0.69	49.22	46.48	0.69	0.69
Fold5	1	63.28	53.52	5.43	7.40	46.88	53.52	7.97	7.40
	10	48.44	46.48	0.69	0.69	40.63	46.48	0.69	0.69
	20	50.00	46.48	0.69	0.69	45.31	46.48	0.69	0.69
	30	49.22	46.48	0.69	0.69	46.09	46.48	0.69	0.69
	40	50.00	46.48	0.69	0.69	53.13	46.48	0.69	0.69

ing all the slices of structural MR to one dataset, and the slices with a clear and complete appearance of gray and white matter to another dataset. In order to create the ADHD classification model, known CNN architectures such as AlexNet, VggNet, ResNet, SqueezeNet, as well as our own CNN network were used. The performance evaluation of the models was made using the 5 Fold Cross Validation method. A classification performance above 85% was achieved with the AlexNet architecture and the dataset containing all structural MR sections. In addition, a classification performance above 85% was achieved with our lesser layered CNN network, which we designed based on the AlexNet architecture. The results show that higher classification performances are achieved in deep learning architectures when the information that defines the data of the problem is given exactly. With a few sections of struc-

Table 9: Performance results of augmented data with recommended CNN network.

		DataSet-1 [Specific Slices]				DataSet-2 [All Slices]			
Fold	Epoch	T Acc.	V Acc.	T Loss.	V Loss.	T Acc.	V Acc.	T Loss.	V Loss.
Fold1	1	47.89	52.11	5.18	7.64	46.48	52.11	3.65	7.63
	10	52.11	60.56	0.58	0.53	77.46	95.77	0.43	0.08
	20	54.93	60.56	0.57	0.54	**76.06**	**95.77**	**0.29**	**0.07**
	30	56.34	60.56	0.58	0.54	**84.51**	**95.77**	**0.22**	**0.07**
	40	53.52	70.42	0.57	0.52	**87.32**	**95.77**	**0.18**	**0.07**
Fold2	1	53.52	52.11	4.38	7.63	56.34	56.34	3.77	6.44
	10	71.83	76.06	0.58	0.55	**80.28**	**97.18**	**0.24**	**0.05**
	20	67.61	76.06	0.61	0.58	**88.73**	**98.59**	**0.17**	**0.04**
	30	67.61	76.06	0.57	0.54	**85.92**	**98.59**	**0.19**	**0.04**
	40	71.83	76.06	0.54	0.51	**87.32**	**98.59**	**0.16**	**0.04**
Fold3	1	56.34	52.11	3.93	7.63	56.34	54.93	2.43	6.45
	10	66.20	71.83	0.61	0.57	**81.69**	**94.37**	**0.30**	**0.12**
	20	64.79	73.24	0.58	0.55	**83.10**	**95.77**	**0.21**	**0.07**
	30	70.42	73.24	0.56	0.53	**83.10**	**95.77**	**0.23**	**0.06**
	40	61.97	73.24	0.53	0.51	**84.51**	**95.77**	**0.21**	**0.05**
Fold4	1	45.07	46.48	3.57	8.53	43.66	53.52	4.10	7.40
	10	40.85	53.52	0.69	0.69	**73.24**	**92.96**	**0.46**	**0.46**
	20	46.48	53.52	0.69	0.69	**83.10**	**95.77**	**0.47**	**0.43**
	30	50.15	53.52	0.69	0.69	**85.92**	**98.59**	**0.34**	**0.24**
	40	59.15	53.52	0.69	0.69	**90.14**	**98.59**	**0.27**	**0.14**
Fold5	1	46.48	53.52	4.60	7.40	54.93	59.15	2.37	2.39
	10	71.83	80.28	0.53	0.45	**81.69**	**95.77**	**0.28**	**0.07**
	20	77.46	81.69	0.45	0.40	**85.92**	**100.0**	**0.22**	**0.02**
	30	70.42	81.69	0.46	0.38	**87.32**	**100.0**	**0.81**	**0.02**
	40	70.42	81.69	0.48	0.37	**85.92**	**100.0**	**0.17**	**0.01**

tural MR data, it shows that the data describing the problem is missing and the classification performance gives poorer results compared to the other dataset.

Conflicts of Interest

All authors; Gulay Cicek and Aydin Akan declare that they do not have conflict of interest.

Funding and Acknowledgement

This work was supported by Izmir Katip Celebi University Scientific Research Projects Coordination Unit: Project numbers 2019-GAP-MÜMF-003 and 2017-ÖNAP-MÜMF-0002.

References

Alex Krizhevsky, Ilya Sutskever and Geoffrey E. Hinton. (2012). Imagenet classification with deep convolutional neural networks. *Advances in Neural Information Processing Systems*, 25: 1097–1105.

Amirmasoud Ahmadi, Mehrdad Kashefi, Hassan Shahrokhi and Mohammad Ali Nazari. (2021). Computer aided diagnosis system using deep convolutional neural networks for adhd subtypes. *Biomedical Signal Processing and Control*, 63: 102–227.

Erik Andersson and Robin Berglund. (2018). Evaluation of data augmentation of mr images for deep learning.

Guilherme V. Polanczyk, Giovanni A. Salum, Luisa S. Sugaya, Arthur Caye, Luis A. Rohde et al. (2015). Annual research review: A meta-analysis of the worldwide prevalence of mental disorders in children and adolescents. *Journal of Child Psychology and Psychiatry*, 56(3): 345–365.

John Murphy. (2016). An overview of convolutional neural network architectures for deep learning. *Microway Inc.*

Li Zhu, Liying Zhang, Yuntao Han, Quan Zeng and Weike Chang. (2017). Study of attention deficit/hyperactivity disorder classification based on convolutional neural networks. *Sheng wu yi xue gong cheng xue za zhi= Journal of biomedical engineering= Shengwu yixue gongchengxue zazhi*, 34(1): 99–105.

Liang Zou, Jiannan Zheng, Chunyan Miao, Martin J. Mckeown and Jane Z. Wang. (2017). 3d cnn based automatic diagnosis of attention deficit hyperactivity disorder using functional and structural mri. *IEEE Access*, 5: 23626–23636.

Lizhen Shao, Donghui Zhang, Haipeng Du and Dongmei Fu. (2019). Deep forest in adhd data classification. *IEEE Access*, 7: 137913–137919.

Muhammad Naveed Iqbal Qureshi, Jooyoung Oh, Beomjun Min, Hang Joon Jo, Boreom Lee et al. (2017). Multi-modal, multi-measure, and multiclass discrimination of adhd with hierarchical feature extraction and extreme learning machine using structural and functional brain mri. *Frontiers in Human Neuroscience*, 11: 157.

Muneeb ul Hassan. (2018). Vgg16-convolutional network for classification and detection. *en línea]. [consulta: 10 abril 2019]. Disponible en: https://neurohive. io/en/popular-networks/vgg16*.

Özlem Kahraman and Esra Özdemir Demirci. (2018). Internet addiction and attention-deficit–hyperactivity disorder: Effects of anxiety, depression and self-esteem. *Pediatrics International*, 60(6): 529–534.

Rashid Amin, Mohammed A. Al Ghamdi, Sultan H. Almotiri and Meshrif Alruily. (2021). Healthcare techniques through deep learning: Issues, challenges and opportunities. *IEEE Access*, 9: 98523–98541.

Robert Rader, Larry McCauley and Erin C. Callen. (2009). Current strategies in the diagnosis and treatment of childhood attention-deficit/hyperactivity disorder. *American Family Physician*, 79(8): 657–665.

Robert Woodard. (2006). The diagnosis and medical treatment of adhd in children and adolescents in primary care: A practical guide. *Pediatric Nursing*, 32(4).

Tianyi Wang and Sei-ichiro Kamata. (2019). Classification of structural mri images in adhd using 3d fractal dimension complexity map. In *2019 IEEE International Conference on Image Processing (ICIP)*, pp. 215–219. IEEE.

Yanli Zhang-James, Emily C. Helminen, Jinru Liu, Barbara Franke, Martine Hoogman et al. (2019). Machine learning classification of attention-deficit/hyperactivity disorder using structural mri data. *bioRxiv*, pp. 546671.

Yuyang Luo, Tara L. Alvarez, Jeffrey M. Halperin and Xiaobo Li. (2020). Multimodal neuroimaging-based prediction of adult outcomes in childhood-onset adhd using ensemble learning techniques. *NeuroImage: Clinical*, 26: 102–238.

Chapter 3

Basic Ensembles of Vanilla-Style Deep Learning Models Improve Liver Segmentation From CT Images

A. Emre Kavur,[a] Ludmila I. Kuncheva[b] and M. Alper Selver[c],*

ABSTRACT

Segmentation of the liver from 3D computer tomography (CT) images is one of the most frequently performed operations in medical image analysis. In the past decade, Deep Learning Models (DMs) have offered significant improvements over previous methods for liver segmentation. The success of DMs is usually owed to the user's expertise in deep learning as well as to intricate training procedures. The need for bespoke expertise limits the reproducibility of empirical studies involving DMs. Today's consensus is that an ensemble of DMs works better than the individual component DMs. In this study we set off to explore the potential of ensembles of publicly available, 'vanilla style DM segmenters. Our ensembles were created from four off-the-shelf DMs: U-Net, Deepmedic, V-Net, and Dense V-Networks. To prevent further overfitting and to keep the overall model simple, we use basic non-

---

[a] Graduate School of Natural and Applied Sciences, Dokuz Eylul University, Izmir, Turkey.
[b] School of Computer Science and Electronic Engineering, Bangor University, Bangor, United Kingdom.
[c] Department of Electrical and Electronics Engineering, Dokuz Eylul University, Izmir, Turkey.
* Corresponding author: alper.selver@deu.edu.tr

trainable ensemble combiners: majority vote, average, product and min/max. Our results with two publicly available data sets (CHAOS and 3Dircadb1) demonstrate that ensembles are significantly better than the individual segmenters on four widely used metrics.

1 Introduction

Segmentation has been recently reported to be the most studied field of biomedical image processing, accounting for around 70% of all studies Maier-Hein et al. (2018). Liver segmentation is a particularly important topic in medical image segmentation due to its use in numerous clinical procedures including but not limited to volumetry Lu et al. (2017), early-stage diagnosis Moghbel et al. (2018), tumour/lesion detection Li et al. (2018); Chlebus et al. (2018); Christ et al. (2016); Vorontsov et al. (2019), disease classification Bal et al. (2018), surgery, and radiotherapy treatment planning Yang et al. (2018). Among many emerging machine learning paradigms, Deep Learning Models (DMs) and particularly Convolutional Neural Networks (CNNs), have achieved remarkable results Zhou et al. (2017). The two main requirements of DMs are (1) the availability of extensively annotated datasets for training, and (2) user experience and expertise for constructing successful architectures and adjusting their parameters. The vibrant interest in the area of medical image segmentation has prompted the organisation of contests and challenges, whereby common databases are released. Over the last five years, in a typical medical imaging challenge, participants prefer to design a dedicated DM or implement and tune a previously designed DM such as U-Net, Deepmedic, V-Net and Dense V-Networks. Kamnitsas et al. (2016); Ronneberger et al. (2015); Milletari et al. (2016); Gibson et al. (2018a).

Available 3D volumetric data sets contain only a few tens of images Heimann et al. (2009); Bilic et al. (2019); Menze et al. (2015) due to the high expense of gathering and annotating such data sets. This number is far too small for proper training of a DM and could lead to spurious results due to overfitting. Classifier ensembles are known to achieve better results compared to their base classifiers Kuncheva (2014) even when those classifiers are overfitted. Organisers of medical segmentation competitions often demonstrate the ensemble superiority combining the top 5 or so entries in the league table. Kamnitsas et al. (2018); Isensee et al. (2019) the practical use of such ensembles is questionable because of the following. Multiple submissions are normally allowed, whose results are disclosed to the participant. This allows the participant to tune their model on the *testing data*, which is a form of 'peeking' Smialowski et al. (2010); Reunanen (2003); Diciotti et al. (2013). Therefore, an ensemble of contest winners may be over-tuned on the testing data. Besides, the ensemble members are likely to be sophisticated (and not always reproducible) models of their own.

Contrary to most recent studies, we propose, what we call, a 'vanilla'-style ensemble where the DMs are state-of-the-art baseline models with no change of their default parameters. Next, we propose four basic combination methods as ensemble combiners for the following reason. Given the small size of the training data (number of patients), further tuning is not advisable Duin (2002). The vanilla ensemble gives the practitioner a ready-made solution, eliminating the need for elaborate tuning and structure modifications of the model, both of which require bespoke expertise and sometimes just luck.

The rest of the paper is organised as follows. Section 2 summarises the related work on medical image segmentation using DMs and ensembles thereof. Section 3 gives details of the four DMs used in this study, the ensemble combination rules, and the four metrics used to evaluate the segmentation results. Our motivation for using ensembles is illustrated through an example. The experiment with the CHAOS competition data set Kavur et al. (2019c) and the 3Dircadb dataset IRCAD (2009) is reported in Section 4. Finally, Section 5 gives our conclusions and outlines future research directions.

2 Related Work

Deep learning has rightfully attracted widespread attention in the literature by outperforming alternative machine learning approaches in various fields of applications, especially in medical imaging Greenspan et al. (2016); Shin et al. (2016).

Further on, combining multiple results coming from different models to achieve a final refined outcome has become an effective way to obtain superior results Kuncheva (2014). It has been recognised that most high-profile competitions such as Imagenet[1] and Kaggle[2] are won by ensembles of deep learning architectures Huang et al. (2017). Lately, ensembles of DMs have been proposed for various domain applications, for example, object detection Razinkov et al. (2018), video classification Zheng et al. (2019), aerial scene classification Dede et al. (2019), and diagnosis and prediction in industrial systems and processes Ma and Chu (2019); Zhang et al. (2017).

Usually, a small group of DMs are taken as the ensemble members due to high computational costs. The models are trained either separately Kamnitsas et al. (2017); Warfield et al. (2004) or simultaneously, together with another network that combines them Dede et al. (2019); Ma and Chu (2019); Zheng et al. (2019). Some of the ensemble methods train *different* DMs (heterogeneous

---
[1] http://www.image-net.org/challenges/LSVRC/.
[2] https://www.kaggle.com/competitions.

ensembles) while others train the same DM with different parameters, training patterns or data. An interesting training strategy that avoids training multiple models is "snapshot ensembling" Huang et al. (2017), whereby the training process of a single DM is stopped at different local minima, and the respective DM is retrieved. The ensemble is then constructed from these (possibly undertrained but supposedly diverse) DMs.

For individually trained DMs, simple ensemble combination rules are applied such as majority vote Ortiz et al. (2016), average Kamnitsas et al. (2018); Maji et al. (2016); Codella et al. (2017), product and more. Warfield et al. (2004); Ju et al. (2018). Warfield et al. (2004) propose an expectation maximisation method named STAPLE for the combination of image segmenters.[3] Trained combination rules such as stacked generalisation, Bayes models and "super learner" Ju et al. (2018) have also been considered for this task.

The remarkable potential of DM ensembles in medical imaging is often illustrated at the end of public competitions Prevedello et al. (2019). The top methods (usually DMs) are combined through majority voting, and the result is usually better than the best contestant's result Menze et al. (2015); Jimenez-del-Toro et al. (2016); Bilic et al. (2019); Kamnitsas et al. (2018); Kavur et al. (2019a). While this makes a compelling case for DM ensembles, such results could be misleading. The individual ensemble members (the contestant entries) have been honed on the testing data during the competition, which means that the testing data is no longer independent of the training data, and the result of the ensemble (and the individual members) may be optimistically biased. (The "peeking" phenomenon.)

A multitude of DM ensembles has been proposed specifically for medical image segmentation Ju et al. (2018); Codella et al. (2017). Bilic et al. (2019) published a compelling overview of the results of the Liver Tumour Segmentation Benchmark (LiTS) contest held in 2017.[4] The study analyses the winning state-of-the-art models in the competition. Although the focus is on segmenting tumours in the liver, the segmentation of the liver itself is also considered. Among other conclusions, the authors note the following:

1. It is still difficult to provide recommendations with regards to the exact network design (structure, parameters, modifications, training). The current solutions are mostly guided by rough ideas instead of strict, proven guidelines. Exploring the possible choices for each task at hand is largely hindered by long training times.

---
[3] http://crl.med.harvard.edu/software/.
[4] Organised in conjunction with the IEEE International Symposium on Biomedical Imaging (ISBI) 2017 and the International Conference on Medical Image Computing & Computer Assisted Intervention (MICCAI) 2017.

2. Only a few of the best DMs were fully-3D (that is, taking a volume image as input). This was again attributed to excessive computational costs. The authors reported that potentially, full 3D DMs would be more successful than the currently used sparse 3D or 2.5D DMs. Here we can add a comment that full 3D DMs may need even more parameters to be set up compared to the non-3D models, which makes the practitioner's task harder.
3. It was observed that ensemble methods outperformed in general the single segmenter methods but their practical use may be hampered by computational constraints. In other words, it is questionable whether the gain the ensembles offer justifies the extra computational resources needed. Our experiments here show that ensembles are significantly better than their individual components, and therefore we argue that the extra cost is well justified.

To address the first concern, here we propose to use off-the-shelf (vanilla-style) DMs. This will eliminate the need to tune any parameter, modify the structure or devise a bespoke training protocol. Second, we propose to use full 3D DMs, which are available at this stage, and their training takes a reasonable time. Third, we will demonstrate that the individual segmenters are outperformed by the ensemble segmenters by a large margin, and therefore we advocate the ensemble approach.

We chose to use a heterogeneous ensemble of four 3D DMs because this ensemble type was found to be superior to homogeneous ensembles of DMs, even to those trained by snapshot ensembling Dede et al. (2019). We kept the combination rules as simple as possible for two reasons. First, we eliminate further overfitting of the ensemble, and second, simple non-trainable combiners are straightforward to implement and do not require profound expertise in either DMs or ensemble methods.

3 Methods

3.1 DMs for Liver Segmentation

The four well-established CNNs used as our ensemble members are detailed below.

3.1.1 U-Net

U-Net is one of the first convolutional neural networks designed for the segmentation of biomedical images. Long et al. (2015). U-Net has been designed to operate with fewer training data compared to standard CNNs without compromising the segmentation accuracy. In this work, we chose the original implementation of U-Net from NiftyNet with cross-entropy as the loss function.[5] NiftyNet provides an open-source front-end platform for different CNN solutions Gibson et al. (2018b); Li et al. (2017) for the assessment of medical images. NiftyNet offers modular design so that different CNNs can be constructed.

3.1.2 Deepmedic

DeepMedic is a multi-scaled 3D Deep Convolutional Neural Network combined with a linked 3D fully connected Random Field Kamnitsas et al. (2016). Deepmedic was originally designed for brain lesion segmentation and won the ISLES 2015 and BraTS 2017 challenges Kamnitsas et al. (2017). Deepmedic was used from its original source. The system was directly downloaded and applied to our data.[6]

3.1.3 V-Net

V-Net is designed for volumetric segmentation of the prostate from a MR image series Milletari et al. (2016). The whole pathway has a V-shape, which is where the CNN gets its name. While V-Net and U-Net share similar structures, their loss functions are different. The V-Net model used here was also sourced from NiftyNet.[7]

3.1.4 Dense V-Networks

Dense V-Networks are developed for automatic segmentation of abdominal organs from CT scans Gibson et al. (2018a). They differ from the other DMs by including three dense feature blocks at each encoding stage. They use the probabilistic Dice score as their loss function.

The Dense V-network model used here was taken from NiftyNet.[8]

3.2 Ensemble Combination Methods

We consider the simplest ensemble combination methods which do not require any further training or parameter tuning Kuncheva (2014). Denoted by

---

[5] https://niftynet.readthedocs.io/en/dev/_modules/niftynet/network/unet.html.
[6] https://github.com/deepmedic/deepmedic.
[7] https://niftynet.readthedocs.io/en/dev/_modules/niftynet/network/vnet.html.
[8] https://niftynet.readthedocs.io/en/dev/_modules/niftynet/network/dense_vnet.html.

$p_1, ..., p_4$ the values of the probability maps outputted by the four segmenters for a given voxel in the 3D image. These values estimate the probability that the voxel is from the class foreground. The combination methods in this paper are as follows:

- *Majority voting.* The probability map of each segmenter is thresholded at 0.5. The voxels with values above this threshold are considered foreground (liver) and the others, background. Having four segmenters, each voxel receives four votes. The voxel is labelled as foreground by the Majority Voting method if 3 or 4 votes are for the foreground.

- *Average combiner.* For each voxel, we calculate the support for the class foreground by averaging the four probability map values: $P_{foreground} = \frac{1}{4}(p_1 + p_2 + p_3 + p_4)$. If $P_f > 0.5$, the class foreground is assigned to that voxel. The class background is assigned otherwise.

- *Product combiner.* For each voxel, we calculate the support for class foreground by $P_f = -\log(P_{f0}) + \sum_{1}^{P4} \log(p_i)$, where any base logarithm can be used, and P_{f0} denotes the prior probability of class foreground. This probability can be estimated from the training data as the proportion of foreground voxels in all images. We next calculate the support for class background by $P_b = -\log(1 - P_{f0}) + \sum_{1}^{P4} 1 - \log(p_i)$. We assign class foreground to the voxel if $P_f > P_b$, and class background, otherwise.

- *Min/Max combiner.* The minimum and the maximum combination rules are identical for the case of two classes Kuncheva (2014), which is why we report one value for both. In this combiner, we calculate $P_f = \min_i(p_i)$ and $P_b = \min_i(1 - p_i)$. Again, we assign class foreground to the voxel if $P_f > P_b$, and class background, otherwise.

3.3 Evaluation Metrics

The primary aim of medical image segmentation is to develop tools for clinical needs such as diagnosis, surgery planning, and organ transplant operations. Hence, tolerance of error is minimal. According to previous studies MaierHein et al. (2018); Yeghiazaryan et al. (2015), there is no single metric that evaluates 3D segmented data completely and fairly in terms of clinically acceptable results. Since the results have to be analysed from many perspectives, the aggregation of multiple evaluation metrics was preferred Langville and Meyer (2013). Spatial overlap-based, volume-based, and spatial distance-based metrics were chosen here to analyse different aspects of segmentation in terms of different aspects.

The four metrics used to evaluate performance for a segmentation result are listed below:

1. Dice coefficient (DICE). Denoting the set of foreground voxels in the candidate segmentation by X and that for the ground truth by Y, the Dice coefficient is calculated as Dice = $2|X \cap Y|/(|X| + |Y|)$, where $|.|$ denotes cardinality (the larger, the better).
2. Relative absolute volume difference (RAVD). Using the notation above, RAVD = abs($|X| - |Y|$)/$|Y|$, where 'abs' denotes the absolute value (the smaller, the better).
3. Average symmetric surface distance (ASSD). This metric is the average Hausdorff distance (in millimetre) between border voxels in X and Y (the smaller, the better).
4. Maximum symmetric surface distance (MSSD). This metric is the maximum Hausdorff distance (in millimetre) between border voxels in X and Y (the smaller, the better).

The code for all metrics (in MATLAB®, Python and Julia) are available at *https://github.com/emrekavur/CHAOS-evaluation*, where we also provide the metrics' calculation details and CHAOS scoring system.

3.4 Why are Ensembles Better?

Ensembles of *diverse* segmenters can clear erroneous artefacts and smoothen out spurious contours in the individual segmentations. This is expected to lead to a better overall match to the ground truth.

Figure 1 shows the results of the four individual segmenters. The grey level intensity reflects the probability map of foreground versus background. The white contour is the proposed segmentation boundary obtained by thresholding the probability map at 0.5. The red contour is the ground truth.[9]

In order to show the advantage of the ensemble, we chose a small, notoriously difficult, region to zoom on: *vena cava superior* (the blue rectangle in the U-Net plot in Fig. 1). Figure 2 contains five plots of the segmented region. The product combiner was chosen for the ensemble. The ground truth is shaded in blue in the ensemble plot, and in red in the plots for the individual segmenters. The guessed segmentation is overlaid in transparent grey. The Dice score for the chosen area is shown under each plot. A Dice score of 1 indicates perfect segmentation while lower values indicate a mismatch. As the results show, both visually and through the numbers, the ensemble segmentation is better than any of the individual ones.

---

[9] The image for this example is slide 11 from the data of patient number 4 in the CHAOS dataset.

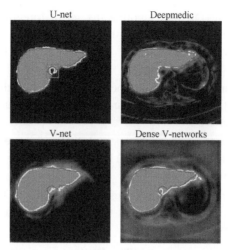

Figure 1: Illustration of the segmentation results of the individual segmenters. White lines represent the border of the segmentation results while black lines represent the border of the ground truth.

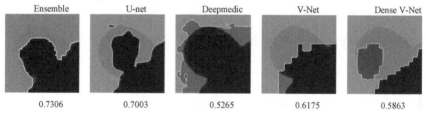

Figure 2: The ensemble segmentation with the product combination rule and the individual segmentations zoomed on *vena cava superior* (the gray rectangle in the U-net plot in Fig. 1). The ground truth is shaded in lighter gray in the ensemble plot, and the plots for the individual segmenters. The guessed segmentation is overlaid in darker gray. The Dice score for the chosen area is shown under each plot.

4 Experimental Evaluation

Of course, one example can serve as an illustration but is not proof. To give empirical support to our claim that vanilla-style ensembles of DMs are superior to the individual DMs, we carry out an experiment with two publicly available datasets.

The purpose of this experiment is twofold. First, we demonstrate the overfitting effect exhibited by DMs on a relatively small data set. The second branch of our experiment explores the four classifier ensemble combiners for liver segmentation from CT scans. Implementation and evaluation of ensemble method codes for a sample patient CT set are available at *https://github.com/emrekavur/Basic-Ensembles-of-DMs-for-Liver-Segmentation* page.

4.1 Data

In order to demonstrate the consistency of the results, we performed our experiments on two publicly available databases that were published at different times and have different characteristics.

- *CHAOS Competition Data*

The data consists of 40 sets of images taken from the Combined (CT-MR) Healthy Abdominal Organ Segmentation Challenge (CHAOS) Kavur et al. (2019b,c). CHAOS is a medical imaging challenge focused on the segmentation of healthy abdominal organs from CT and/or MRI image series Kavur et al. (2020). CHAOS was started during the IEEE International Symposium on Biomedical Imaging (ISBI) on April 11, 2019, in Venice, Italy. CHAOS has five independent tasks related to different modalities and abdominal organs. Participants may join the challenge for a single task or multiple tasks.[10]

In this work, the CT part of the CHAOS data was used. The CT data contains abdomen images of 40 different patients who have healthy livers. The technical information about the data is presented in Table 1.

The training part of the data contains 20 randomly chosen anonymised DICOM images and their ground truths. However, the test data includes only the anonymised DICOM images of the remaining 20 patients. The evaluation of the results on the test data is done by an evaluation script hosted on the grand-challenge.org website. This ensures that CHAOS competitors do not have access to the ground truth (manual segmentation) of the testing data. However, since multiple submissions are allowed and the overall score across the testing data is returned to the participant, the competitors can gauge the

Table 1: Statistics about CHAOS CT and 3DIRCADb datasets.

	3DIRCADb	CHAOS
Number of 3D image sets	20 (10 + 10)	40 (20 + 20)
Spatial resolution of files	512 × 512	512 × 512
Number of files (slices) in all cases [min–max]	[74–260]	[78–294]
Average number of files in the cases	141	160
Total number of files in the dataset	2823	6407
X space (mm) [min–max]	[0.56–0.87]	[0.54–0.79]
Y space (mm) [min–max]	[0.56–0.87]	[0.54–0.79]
Slice thickness (mm) [min–max]	[1.60–4.00]	[2.00–3.20]

---

[10] As of January 2020, CHAOS is one of the most popular medical imaging challenges in grand-challenge.org and has 1516 current participants at the website Kavur et al. (2019b, 2020).

success of their solution and tune it accordingly (the peeking phenomenon). In our experiments, we report results on the testing data, but we do not tune any of our methods on these results.

- *3Dircadb1 Data*

The second dataset, 3D-IRCADb-01 (3D Image Reconstruction for Algorithm Data Comparison) IRCAD (2009) contains abdomen CT scans of 20 patients. Unlike the CHAOS dataset, 75% of cases have hepatic tumours in the liver. 3D-IRCADb not only provides ground truths of livers but also various structures such as all hepatic veins and hepatic tumours. The ground truths were created by clinical experts. Except for tumours, all structures inside the liver were considered together as segmentation targets in this work. We equally divided 3D-IRCADb into two parts for training and testing (10 sets for each). The technical information about each dataset is presented in Table 1. With 3Dircadb1 data, the same procedures were followed with CHAOS data for developing, training, and testing.

4.2 Experimental Protocol

For each dataset, the four DMs were trained on the training data. For an input image (3D CT scan), each DM returns a 3D probability map P. For voxel at (i, j, k), $P(i, j, k)$ is the probability that the voxel belongs to the liver. The output probability maps are then smoothed using $3 \times 3 \times 3$ convolution kernel (function smooth3 in MATLAB) to eliminate small defects in the segmentation. We also smoothed, in the same way, the ensemble output after applying the respective combination rule.

Next, we evaluated the four metrics on the training data and then on the testing data, both for the individual DMs and the four ensembles.

Finally, we tested the hypothesis that ensembles are better than individual DM segmenters on the testing data by running tests for each (ensemble, segmenter) pair, for each of the four metrics.

To determine the statistical significance of the difference between methods A and B, we applied the following protocol:

- *Paired samples.* Suppose that x and y are the vectors containing the values for methods A and B, respectively. (For example, x may contain the Dice scores obtained from Deepmedic for the testing data from 20 patients in the CHAOS data, and y may contain the Dice scores obtained from the Majority vote ensemble for the same 20 patients.) We are interested in whether the means of x and y are significantly different. Using Lilliefors goodness-of-fit test, we check the normality of the difference $x - y$. If normality cannot be rejected at 0.05 level, we use the paired t-test for comparing the means. If the normality of the difference does not hold, we use Wilcoxon signed-rank test for zero medians of the difference.

- *Non-paired samples.* For non-paired samples, we check the normality of x and y. If both hold, we apply a 2-sample t-test. Otherwise, we use the Wilcoxon rank-sum test (Mann–Whitney U test).

4.3 Results

A full set of results (all metrics for all DMs and all ensembles; training and testing) is provided in *https://github.com/emrekavur/Basic-Ensembles-of-DMs-for-Liver-Segmentation* page. Tables 2–5 show the average results for the two datasets, all metrics, individual DMs and ensembles.

Table 2: Metric results on CHAOS training data for the individual segmenters (rows 1–5) and the ensemble methods (rows 6–9). The circle marker indicates results where the overfitting was not found to be significant.

	DICE	RAVD	ASSD	MSSD
U-Net	0.935	∘14.800	3.903	54.650
Deepmedic	0.984	1.115	1.709	67.078
V-Net	0.948	∘3.824	1.656	42.972
Dense V-net	0.932	3.039	2.289	∘78.118
Average	0.950	5.694	2.389	60.705
Majority Vote	0.976	2.401	0.746	11.043
Average	0.981	1.003	0.637	11.621
Product	0.975	∘3.493	0.888	12.581
Min-Max	0.978	1.208	0.811	11.559

Table 3: Metric results on CHAOS test data for the individual segmenters (rows 1–5) and the ensemble methods (rows 6–9). The best value in each column is enclosed in a box.

	DICE	RAVD	ASSD	MSSD
U-Net	0.811	54.842	14.253	104.515
Deepmedic	0.951	3.058	7.174	141.473
V-Net	0.879	|17.434|	6.146	104.189
Dense V-net	0.886	7.702	4.492	113.139
Average	0.882	20.759	8.016	115.829
Majority Vote	0.952	4.235	1.719	28.517
Average	0.953	3.839	|1.956|	|30.676|
Product	|0.946|	6.867	2.121	32.696
Min-Max	0.937	6.094	2.311	35.052

Table 4: Metric results on 3Dircadb1 training data for the individual segmenters (rows 1–5) and the ensemble methods (rows 6–9). The circle marker indicates results where the overfitting was not found to be significant.

	DICE	RAVD	ASSD	MSSD
U-Net	0.904	19.320	∘7.075	∘70.798
Deepmedic	0.988	0.216	0.399	36.127
V-Net	0.968	2.974	0.975	17.587
Dense V-net	0.973	1.110	0.930	∘51.780
Average	0.958	5.905	2.345	44.073
Majority Vote	0.979	2.432	0.633	12.913
Average	0.982	1.844	0.615	18.759
Product	0.978	∘3.426	0.841	20.764
Min-Max	0.980	2.052	0.712	19.423

Table 5: Metric results on 3Dircadb1 test data for the individual segmenters (rows 1–5) and the ensemble methods (rows 6–9). The best value in each column is enclosed in a box.

	DICE	RAVD	ASSD	MSSD
U-Net	0.672	75.092	66.551	172.201
Deepmedic	0.905	10.385	4.753	139.120
V-Net	0.828	19.182	8.913	95.328
Dense V-net	0.902	8.726	9.009	104.886
Average	0.827	28.346	22.306	127.884
Majority Vote	0.890	14.348	3.341	|55.303|
Average	|0.920|	7.131	|3.070|	74.613
Product	0.916	|6.418|	3.271	73.580
Min-Max	0.906	8.799	3.790	76.629

4.3.1 Ensemble Segmenters Show Less Overfitting than Individual DMs

The overfitting can be observed from the tables. Without exception, both for individual DMs and ensembles, the average training values are preferable to the averaged testing values. In some cases, the differences are substantial, for example, U-Net gives DICE = 0.904 for the training data and 0.672 for the testing data. Even though the large differences in the metrics' values suggest otherwise, further tests revealed that 8 of the 64 differences are not statistically significant at a level of 0.05. The 8 results are marked with a circle marker in Tables 2 and 4.

The results thus far show that while ensembles reach better metric values they are not immune to overfitting. Nonetheless, we will show that, in general, they exhibit less overfitting compared to the individual DMs. To visualise this, we present two Tables (6 and 7), one for each dataset, which contain the overfitting magnitude calculated as the training value minus the testing value. For DICE, a positive difference means that the training value was better. For the other three metrics, negative values indicate that the training value was better because lower values of these metrics are preferable. The values for each metric are colour-coded. Red colours indicate smaller overfitting while blue colours indicate larger overfitting. The blue colour is present more in the top parts of both tables showing that the individual segmenters are more prone to that than the ensembles.

4.3.2 Ensemble Segmenters Offer Better Results than Individual DMs

Tables 3 and 5 show that ensemble values are, in most part, preferable to the values of the individual DMs on all four metrics. We compared the individual DMs and the ensembles on the testing data using our statistical protocol for paired data. Tables 8–11 details the results from the statistical comparisons.

Table 6: Overfitting magnitude for the CHAOS dataset. Table clearly shows that overfitting occurs less in ensemble methods.

	DICE	RAVD	ASSD	MSSD
U-Net	0.1238	40.0423	10.3499	49.8649
Deepmedic	0.0329	−1.9436	−5.4651	74.3951
V-Net	0.0695	13.6101	−4.4899	61.2170
Dense V-networks	0.0463	−4.6631	−2.2030	35.0205
Majority	0.0237	−1.8338	−0.9726	17.4741
Average	0.0268	−2.8366	−1.3186	19.0548
Product	0.0281	−3.3736	−1.2328	20.1146
Min-Max	0.0381	−4.8866	−1.4997	23.4931

Table 7: Overfitting magnitude for the 3Dircadb1 dataset. Table clearly shows that overfitting occurs less in ensemble methods.

	DICE	RAVD	ASSD	MSSD
U-Net	0.2320	−55.7723	−59.4755	−101.4027
Deepmedic	0.0830	−10.1691	−4.3534	−102.9930
V-Net	0.1406	−16.2087	−7.9385	−77.7407
Dense V-networks	0.0714	−7.6161	−8.0785	−53.1059
Majority	0.0896	−11.9168	−2.7082	−42.3905
Average	0.0626	−5.2867	−2.4555	−55.8543
Product	0.0614	−2.9921	−2.4300	−52.8161
Min-Max	0.0738	−6.7472	−3.0785	−57.2058

Table 8: Statistical comparison between individual DMs and ensembles for DICE metric. Bullet means that the ensemble wins; circle means that the DM wins; line means that there is no statistical difference.

	CHAOS dataset			
	Majority	Average	Product	Min/Max
U-Net	•	•	•	•
Deepmedic	–	–	–	○
V-Net	•	•	•	•
Dense V	•	•	•	•
	3Dircadb1 dataset			
	Majority	Average	Product	Min/Max
U-Net	•	•	•	•
Deepmedic	–	–	–	–
V-Net	•	•	•	•
Dense V	–	–	–	–

Table 9: Statistical comparison between individual DMs and ensembles for RAVD metric. Bullet means that the ensemble wins; circle means that the DM wins; line means that there is no statistical difference.

	CHAOS dataset			
	Majority	Average	Product	Min/Max
U-Net	•	•	•	•
Deepmedic	○	–	○	–
V-Net	–	•	–	–
Dense V	–	•	–	–
	3Dircadb1 dataset			
	Majority	Average	Product	Min/Max
U-Net	•	•	•	•
Deepmedic	–	–	–	–
V-Net	–	•	•	•
Dense V	–	–	–	–

The level of significance was set everywhere at 0.05. The tables demonstrate that the ensemble segmenters are better than the individual DMs.

Finally, to be able to recommend one of the ensemble models, we present two glyph plots in Figs. 3 and 4. The plots are based on the averaged testing results for each metric. DICE was reversed so that small values are more desirable. The ensemble scores for each metric were scaled between 0.1 and 1 and plotted on the spokes of the glyph plot. An ideal ensemble would occupy a small square in the middle. The larger the surface of the figure presented by the ensemble, the worse the ensemble is in comparison with the rest. The Chaos dataset figure elects the Majority Vote ensemble as the best, closely followed by the Average ensemble. On the other hand, the Majority Vote ensemble

Table 10: Statistical comparison between individual DMs and ensembles for ASSD metric. Bullet means that the ensemble wins; circle means that the DM wins; line means that there is no statistical difference.

CHAOS dataset				
	Majority	Average	Product	Min/Max
U-Net	•	•	•	•
Deepmedic	•	•	•	•
V-Net	•	•	•	•
Dense V	•	•	•	•
3Dircadb1 dataset				
	Majority	Average	Product	Min/Max
U-Net	•	•	•	•
Deepmedic	–	•	•	–
V-Net	•	•	•	•
Dense V	–	•	–	–

Table 11: Statistical comparison between individual DMs and ensembles for MSSD metric. Bullet means that the ensemble wins; circle means that the DM wins; line means that there is no statistical difference.

CHAOS dataset				
	Majority	Average	Product	Min/Max
U-Net	•	•	•	•
Deepmedic	•	•	•	•
V-Net	•	•	•	•
Dense V	•	•	•	•
3Dircadb1 dataset				
	Majority	Average	Product	Min/Max
U-Net	•	•	•	•
Deepmedic	•	•	•	•
V-Net	•	–	–	–
Dense V	•	•	•	•

occupies a large area in the glyph plot in Fig. 4. The Average ensemble is the best for this data set.

Another indication in favour of the Average ensemble is the total number of wins across the data sets, the metrics and the DMs (Tables 8–11). If we add one point for each win and subtract one point for each loss, the ensembles receive the following scores: Majority Vote 20, Average 25, Product 21, and Min/Max 20. (All scores are out of 2 (data sets) × 4 (metrics) × 4 (DMs) = 32 possible points.) This gives us ground to recommend the Average combiner for future use with our vanilla-style ensemble of DMs for liver segmentation.

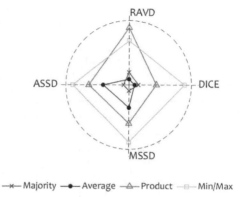

Figure 3: Glyph plot of the four ensemble methods for the CHAOS dataset. The spokes are the four metrics. Small-area ensembles are preferable.

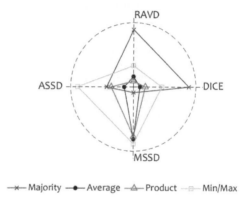

Figure 4: Glyph plot of the four ensemble methods for the 3Dircadb1 dataset. The spokes are the four metrics. Small-area ensembles are preferable.

5 Conclusion

Intrigued by the success of DMs in medical segmentation, we set off to explore the potential of an ensemble of DM segmenters which is composed of publicly available, state-of-the-art DMs, and which does not require profound expertise in deep learning or ensemble methods from the user. The first problem we encountered was that the individual DMs (trained with their default structure, parameters and training setting) are prone to overfitting. Our experimental results confirmed that, and also revealed that the proposed vanilla-style ensemble is less affected by overfitting. Using two publicly available datasets, we demonstrated that a simple ensemble of off-the-shelf deep learning models outperforms the individual ensemble members. The overall message of this study is that it is possible to achieve results in liver segmentation from CT images with minimal programmatic effort, using state-of-the-art DMs as black boxes and basic classifier ensemble rules. Besides, since data/problem-specific design and parameter tuning are not required, the ensemble methods shown in this study can be offered as a solution to the repeatability and reproducibility problems widely seen in deep learning studies. Out of the four combination rules we examined, the average (or sum) combiner achieved the best result.

In addition to the chosen simple combiners, we experimented with several trained combiners: weighted majority vote, Naive Bayes and Behaviour Knowledge Space (BKS) Kuncheva (2014). The results were on par with the simple combiners. This reinforces our message that for small data sizes as the currently available annotated data for liver segmentation, overfitting is a major issue.

By and large, the combination methods recommended in the literature have been evaluated on *classification accuracy*. Here we note that our evaluation hinges on four different metrics. They are not straightforwardly related to classification accuracy. This suggests that developing bespoke training protocols for the DMs as well as more suitable combination strategies could be a good way forward.

We note that the success of the combiners which rely on continuous-valued outputs (Average, Product and Min/Max) critically depends on the calibration of the output of the segmenters. Traditionally, ensembles are constructed with the same base classifier (segmenter here) with different training data, which practically eliminates the problem of calibration. However, if different models are used, as in this study, it is vital to ensure that the probability map calibration is suitable. For example, if one segmenter is very "certain" in its decision and always gives values close to 0 and 1, this segmenter will have a heavier vote. The segmenter may be arbitrarily accurate and may not deserve the advantage over the rest of the segmenters. In this study, we were striving for simplicity and did not calibrate further the four DMs outputs. On the other hand, Dede et al (2019) observe that heterogeneous ensembles of DMs fare better than homogeneous ones, which the authors attribute to the importance of diversity offered by different DM models. Thus, it may pay off to devise and include a calibration pre-processing step in the ensemble pipeline.

Acknowledgements

This work is supported by Scientific and Technological Research Council of Turkey (TUBITAK) with 2214 International Doctoral Research Fellowship Programme and TUBITAK ARDEB-EEEAG under grant number 116E133. The authors gratefully acknowledge the support of NVIDIA Corporation with the donation of the Titan Xp Pascal GPU used for this research.

References

Bal, E., Klang, E. and Hayit Greenspan, M. A. (2018). Automatic liver volume segmentation and fibrosis classification. *Proceedings Volume 10575, Medical Imaging 2018: Computer-Aided Diagnosis 10575*. URL: https://doi.org/10.1117/12.2294555, doi:10.1117/12.2294555.

Bilic, P., Christ, P. F., Vorontsov, E., Chlebus, G., Chen, H. et al. (2019). The liver tumor segmentation benchmark (lits). *CoRR abs/1901.04056*. URL: http://arxiv.org/abs/1901.04056, arXiv:1901.04056.

Chlebus, G., Schenk, A., Moltz, J. H., van Ginneken, B., Hahn, H. K. et al. (2018). Automatic liver tumor segmentation in ct with fully convolutional neural networks and object-based postprocessing. *Scientific Reports*, 8: 15497. URL: https://doi.org/10.1038/s41598-018-33860-7, doi:10.1038/s41598-018-33860-7.

Christ, P. F., Elshaer, M. E. A., Ettlinger, F., Tatavarty, S., Bickel, M. et al. (2016). Automatic liver and lesion segmentation in ct using cascaded fully convolutional neural networks and 3d conditional random fields. pp. 415–423. *In*: Ourselin, S., Joskowicz, L., Sabuncu, M. R., Unal, G. and Wells, W. (eds.). *Medical Image Computing and Computer-Assisted Intervention—MICCAI 2016*, Springer International Publishing, Cham.

Codella, N. C. F., Nguyen, Q., Pankanti, S., Gutman, D. A., Helba, B. et al. (2017). Deep learning ensembles for melanoma recognition in dermoscopy images. *IBM Journal of Research and Development*, 61: 5:1–5:15. doi:10.1147/JRD.2017.2708299.

Dede, M. A., Aptoula, E. and Genc, Y. (2019). Deep network ensembles for aerial scene classification. *IEEE Geoscience and Remote Sensing Letters*, 16: 732–735. doi:10.1109/LGRS.2018.2880136.

Diciotti, S., Ciulli, S., Mascalchi, M., Giannelli, M., Toschi, N. et al. (2013). The "peeking" effect in supervised feature selection on diffusion tensor imaging data. *American Journal of Neuroradiology* URL: http://www.ajnr.org/content/early/2013/07/18/ajnr.A3685, arXiv:http://www.ajnr.org/content/early/2013/07/18/ajnr.A3685.full.pdf.

Duin, R. P. W. (2002). The combining classifier: to train or not to train? *In*: *Proc. 16th International Conference on Pattern Recognition, ICPR'02*, Canada. pp. 765–770.

Gibson, E., Giganti, F., Hu, Y., Bonmati, E., Bandula, S. et al. (2018a). Automatic multi-organ segmentation on abdominal CT with dense V-networks. *IEEE Transactions on Medical Imaging*, 37: 1822–1834. URL: https://ieeexplore.ieee.org/document/8291609/, doi:10.1109/TMI.2018.2806309.

Gibson, E., Li, W., Sudre, C., Fidon, L., Shakir, D. I. et al. (2018b). NiftyNet: A deep-learning platform for medical imaging. *Computer Methods and Programs in Biomedicine*, 158: 113–122. URL: https://www.sciencedirect.com/science/article/pii/S0169260717311823, doi:10.1016/j.cmpb.2018.01.025.

Greenspan, H., van Ginneken, B. and Summers, R. M. (2016). Guest editorial deep learning in medical imaging: Overview and future promise of an exciting new technique. *IEEE Transactions on Medical Imaging*, 35: 1153–1159. doi:10.1109/TMI.2016.2553401.

Heimann, T., van Ginneken, B. and Styner, M. (2009). Comparison and evaluation of methods for liver segmentation from CT datasets. *IEEE Transactions on Medical Imaging*, 28: 1251–1265. URL: http://ieeexplore.ieee.org/document/4781564/, doi:10.1109/TMI.2009.2013851.

Huang, G., Li, Y., Pleiss, G., Liu, Z., Hopcroft, J. E. et al. (2017). Snapshot ensembles: Train 1, get M for free. *CoRR abs/1704.00109*. URL: http://arxiv.org/abs/1704.00109, arXiv:1704.00109.

IRCAD. (2009). 3D-IRCADb (3D Image Reconstruction for Comparison of Algorithm Database). URL: https://www.ircad.fr/research/ 3d-ircadb-01/. accessed: 2019-11-01.

Isensee, F., Petersen, J., Klein, A., Zimmerer, D., Jaeger, P. F. et al. (2019). nnU-Net: Self-adapting framework for U-net-based medical image segmentation. *In*: Informatik aktuell, Springer Vieweg, Wiesbaden. p. 22. URL: http://link.springer.com/10.1007/978-3-658-25326-4_7, doi:10.1007/978-3-658-25326-4_7.

Jimenez-del-Toro, O., Muller, H., Krenn, M., Gruenberg, K., Taha, A. A. et al. (2016). Cloud-based evaluation of anatomical structure segmentation and landmark detection algorithms: Visceral anatomy benchmarks. *IEEE Transactions on Medical Imaging*, 35: 2459–2475. doi:10.1109/TMI.2016.2578680.

Ju, C., Bibaut, A. and van der Laan, M. (2018). The relative performance of ensemble methods with deep convolutional neural networks for image classification. *Journal of Applied Statistics*, 45: 2800–2818. URL: https://doi.org/10.1080/02664763.2018.1441383, doi:10.1080/02664763.2018.1441383.

Kamnitsas, K., Bai, W., Ferrante, E., McDonagh, S., Sinclair, M. et al. (2018). Ensembles of multiple models and architectures for robust brain tumour segmentation. pp. 450–462. *In*: Crimi, A., Bakas, S., Kuijf, H., Menze, B. and Reyes, M. (eds.). *Brainlesion: Glioma, Multiple Sclerosis, Stroke and Traumatic Brain Injuries*, Springer International Publishing, Cham.

Kamnitsas, K., Ferrante, E., Parisot, S., Ledig, C., Nori, A. V. et al. (2016). DeepMedic for brain tumor segmentation. *In*: Lecture Notes in Computer Science (including subseries Lecture Notes in Artificial Intelligence and Lecture Notes in Bioinformatics), Springer, Cham. pp. 138–149. URL: https://link.springer.com/chapter/10.1007/978-3-319-55524-9_14, doi:10.1007/978-3-319-55524-9_14.

Kamnitsas, K., Ledig, C., Newcombe, V. F., Simpson, J. P., Kane, A. D. et al. (2017). Efficient multi-scale 3D CNN with fully connected CRF for accurate brain lesion segmentation. *Medical Image Analysis*, 36: 61–78. URL: https://www. sciencedirect.com/science/article/pii/S1361841516301839, doi:10.1016/j.media.2016.10.004.

Kavur, A. E., Gezer, N. S., Barış, M., Conze, P. H., Groza, V. et al. (2020). CHAOS challenge—combined (CT-MR) healthy abdominal organ segmentation. *arXiv pre-print*. URL: https://arxiv.org/abs/2001.06535, arXiv:2001.06535.

Kavur, A. E., Gezer, N. S., Barış, M., Şahin, Y., Şavas, O. et al. (2019a). Comparison of semi-automatic and deep learning-based automatic methods for liver segmentation in living liver transplant donors. *Diagnostic and Interventional Radiology*, doi:10.5152/dir.2019.19025.

Kavur, A. E., Selver, M. A., Dicle, O., Barış, M., Gezer, N. et al. (2019b). *CHAOS—Combined (CT-MR) Healthy Abdominal Organ Segmentation Challenge—Grand Challenge*. URL: https://chaos.grand-challenge.org/. accessed: 2019-09-12.

Kavur, A. E., Selver, M. A., Dicle, O., Barış, M., Gezer, N. S. et al. (2019c). *CHAOS—Combined (CT-MR) Healthy Abdominal Organ Segmentation Challenge Data*. URL: http://doi.org/10.5281/zenodo.3362844, doi:10.5281/zenodo.3362844. accessed: 2019-04-11.

Kuncheva, L. I. (2014). *Combining Pattern Classifiers: Methods and Algorithms: Second Edition*. Wiley-Interscience. volume 9781118315. pp. 1–357. doi:10.1002/9781118914564.

Langville, A. N. and Meyer, C. D. C. D. (2013). Who's #1? *The Science of Rating and Ranking*. Princeton University Press. p. 247. URL: https://dl.acm.org/citation.cfm?id=2613650.

Li, W., Wang, G., Fidon, L., Ourselin, S., Cardoso, M. J. et al. (2017). On the compactness, efficiency, and representation of 3D convolutional networks: Brain parcellation as a pretext task. In: *Lecture Notes in Computer Science (Including Subseries Lecture Notes in Artificial Intelligence and Lecture Notes in Bioinformatics)*, Springer, Cham. pp. 348–360. URL: http://link.springer.com/10.1007/978-3-319-59050-9_28, doi:10.1007/978-3-319-59050-9_28.

Li, X., Chen, H., Qi, X., Dou, Q., Fu, C. et al. (2018). H-denseunet: Hybrid densely connected unet for liver and tumor segmentation from CT volumes. *IEEE Transactions on Medical Imaging*, 37: 2663–2674. doi:10.1109/TMI.2018.2845918.

Long, J., Shelhamer, E. and Darrell, T. (2015). Fully convolutional networks for semantic segmentation. In: *2015 IEEE Conference on Computer Vision and Pattern Recognition (CVPR)*, IEEE. pp. 3431–3440. URL: http://ieeexplore.ieee.org/document/7298965/, doi:10.1109/CVPR.2015.7298965.

Lu, F., Wu, F., Hu, P., Peng, Z., Kong, D. et al. (2017). Automatic 3d liver location and segmentation via convolutional neural network and graph cut. *International Journal of Computer Assisted Radiology and Surgery*, 12: 171–182. URL: https://doi.org/10.1007/s11548-016-1467-3, doi:10.1007/s11548-016-1467-3.

Ma, S. and Chu, F. (2019). Ensemble deep learning-based fault diagnosis of rotor bearing systems. *Computers in Industry*, 105: 143–152. URL: http://www.sciencedirect.com/science/article/pii/S0166361518304731, doi:https://doi.org/10.1016/j.compind.2018.12.012.

Maier-Hein, L., Eisenmann, M., Reinke, A., Onogur, S., Stankovic, M. et al. (2018). Why rankings of biomedical image analysis competitions should be interpreted with care. *Nature Communications*, 9: 5217. URL: http://www.nature.com/articles/s41467-018-07619-7, doi:10.1038/s41467-018-07619-7.

Maji, D., Santara, A., Mitra, P. and Sheet, D. (2016). Ensemble of deep convolutional neural networks for learning to detect retinal vessels in fundus images. *CoRR abs/1603.04833*. URL: http://arxiv.org/abs/1603.04833, arXiv:1603.04833.

Menze, B. H., Jakab, A., Bauer, S., Kalpathy-Cramer, J., Farahani, K. et al. (2015). The multimodal brain tumor image segmentation benchmark (brats). *IEEE Transactions on Medical Imaging*, 34: 1993–2024. doi:10.1109/TMI.2014.2377694.

Milletari, F., Navab, N. and Ahmadi, S. A. (2016). V-Net: Fully convolutional neural networks for volumetric medical image segmentation. *In: Proceedings—2016 4th International Conference on 3D Vision*, 3DV 2016, IEEE. pp. 565–571. URL: http://ieeexplore.ieee.org/ document/7785132/, doi:10.1109/3DV.2016.79.

Moghbel, M., Mashohor, S., Mahmud, R. and Saripan, M. I. (2018). Review of liver segmentation and computer assisted detection/diagnosis methods in computed tomography. *Artif. Intell. Rev.*, 50: 497–537. URL: https://doi.org/10.1007/s10462-017-9550-x, doi:10.1007/s10462-017-9550-x.

Ortiz, A., Munilla, J., Gorriz, J. M. and Ramírez, J. (2016). Ensembles of deep learning architectures for the early diagnosis of the alzheimer's disease. *International Journal of Neural Systems*, 26: 1650025. URL: https://doi.org/10.1142/S0129065716500258, doi:10.1142/S0129065716500258, arXiv:https://doi.org/10.1142/S0129065716500258. pMID: 27478060.

Prevedello, L. M., Halabi, S. S., Shih, G., Wu, C. C., Kohli, M. D. et al. (2019). Challenges related to artificial intelligence research in medical imaging and the importance of image analysis competitions. *Radiology: Artificial Intelligence*, 1: e180031. URL: https://doi.org/10.1148/ryai.2019180031, doi:10.1148/ryai.2019180031, arXiv:https://doi.org/10.1148/ryai.2019180031.

Razinkov, E., Saveleva, I. and Matas, J. (2018). Alfa: Agglomerative late fusion algorithm for object detection. *In: 2018 24th International Conference on Pattern Recognition (ICPR)*, pp. 2594–2599. doi:10.1109/ICPR.2018.8545182.

Reunanen, J. (2003). Overfitting in making comparisons between variable selection methods. *Journal of Machine Learning Research*, 3: 1371–1382.

Ronneberger, O., Fischer, P. and Brox, T. (2015). U-net: Convolutional networks for biomedical image segmentation. *In: Lecture Notes in Computer Science (including subseries Lecture Notes in Artificial Intelligence and Lecture Notes in Bioinformatics)*, Springer, Cham. pp. 234–241. URL: http://link.springer.com/10.1007/978-3-319-24574-4_28, doi:10.1007/978-3-319-24574-4_28.

Shin, H., Roth, H. R., Gao, M., Lu, L., Xu, Z. et al. (2016). Deep convolutional neural networks for computer-aided detection: Cnn architectures, dataset characteristics and transfer learning. *IEEE Transactions on Medical Imaging*, 35: 1285–1298. doi:10.1109/TMI.2016.2528162.

Smialowski, P., Frishman, D. and Kramer, S. (2010). Pitfalls of supervised feature selection. *Bioinformatics*, 26: 440–443.

Vorontsov, E., Cerny, M., Regnier, P., Di Jorio, L., Pal, C. J. et al. (2019). Deep learning for automated segmentation of liver lesions at ct in patients with colorectal cancer liver metastases. *Radiology: Artificial Intelligence*, 1: 180014. URL: https://doi.org/10.1148/ryai.2019180014, doi:10.1148/ryai.2019180014.

Warfield, S. K., Zou, K. H. and Wells, W. M. (2004). Simultaneous truth and performance level estimation (staple): An algorithm for the validation of image segmentation. *IEEE Transactions on Medical Imaging*, 23: 903–921. doi:10.1109/TMI.2004.828354.

Yang, X., Yang, J. D., Hwang, H. P., Yu, H. C., Ahn, S. et al. (2018). Segmentation of liver and vessels from CT images and classification of liver segments for preoperative liver surgical planning in living donor liver transplantation. *Computer Methods and Programs in Biomedicine*, 158: 41–52. URL: http://www.sciencedirect.com/science/article/pii/S0169260717303383, doi:https://doi.org/10.1016/j.cmpb.2017.12.008.

Yeghiazaryan, V., Voiculescu, I., Yeghiazaryan, V. and Voiculescu, I. (2015). An overview of current evaluation methods used in medical image segmentation. technical report. *Oxford, UK, Department of Computer Science.* URL: https://www.cs.ox.ac.uk/publications/publication10110-abstract.html.

Zhang, C., Lim, P., Qin, A. K. and Tan, K. C. (2017). Multiobjective deep belief networks ensemble for remaining useful life estimation in prognostics. *IEEE Transactions on Neural Networks and Learning Systems*, 28: 2306–2318. doi:10.1109/TNNLS.2016.2582798.

Zheng, J., Cao, X., Zhang, B., Zhen, X., Su, X. et al. (2019). Deep ensemble machine for video classification. *IEEE Transactions on Neural Networks and Learning Systems*, 30: 553–565. doi:10.1109/TNNLS.2018.2844464.

Zhou, S., Greenspan, H. and Shen, D. (2017). *Deep Learning for Medical Image Analysis.* Elsevier Inc.. Chapter 8. pp. 188–191.

Chapter 4

Convolutional Neural Networks for Medical Image Analysis

Rajesh Gogineni[1,*] and *Ashvini Chaturvedi*[2]

ABSTRACT

The enormous advancement of Deep Learning (DL) algorithms in various fields has been adopted by many perspectives of medical image analysis. In the context of deep learning, convolutional neural networks (CNNs) are widely used models for medical imaging tasks and lead to a huge progress in computer aided diagnosis. CNNs are being profoundly absorbed in the research of medical imaging because of their pertinent features in preserving local image relations and dimensionality reduction. CNNs demonstrate excellent performance in lesion detection and other vision tasks. This chapter presents the modules of convolutional neural networks and their prominent architectures employed for medical image analysis. The commonly used variants of CNNs in medical image analysis including ResNet, GoogleNet,

---

[1] Department of E & C Engineering, Dhanekula Institute of Engineering and Technology, Vijayawada, India - 521 139.
[2] Department of E & C Engineering, National Institute of technology Karnataka, Surathkal, India - 575025.
* Corresponding author: rgogineni9@gmail.com

and fully convolutional neural networks are analyzed. The principal research areas and applications of medical image registration, segmentation, detection and classification are discussed. The overview of the above mentioned tasks in a few of the usual diagnosis areas of the human body such as brain, lungs and eye is presented in this chapter. The crucial challenge in medical image analysis using CNNs is the availability of training data sets. This problem is addressed in this chapter, by providing the existing medical image datasets for different application areas like chest, diabetic retinopathy, lungs and heart. Finally, the chapter is concluded by discussing the challenges associated with CNN architectures and promising future research directions.

1 Introduction

The investigation of Medical images copes with the intensification of basic medical data obtained by means of various sources for the purpose of advanced analysis and eclectic visualization. Medical image analysis pursues the progression of different approaches that assist human experts in the objective and subjective estimation of medical images. Medical imaging has developed into visualization and computer-aided diagnosis methods like early detection, diagnosis, and treatment of cancer.

The broad objective of medical image processing is examining the human body with microscopic and macroscopic imaging modalities. The interpretation and characterization of the acquired data requires sophisticated methods. The collated association among biologists, physicians as well as engineers, physicists and computer professionals in addition to medical expertise is prescribed to construct, contrive and to substantiate the medical systems. The recent improvements in artificial intelligence, machine learning and deep neural networks have induced an acute impact in medical imaging field. The analysis of different human body systems like skeletal, digestive and cardiovascular with the aid of clinical data has become a dominant direction in medical image analysis.

The application of deep learning techniques in medical image processing are evolving ardently in a way that most of the existing problems are significantly addressed. The changes in deep learning approaches before and after the introduction of neural networks are presented in Fig. 1. The conventional algorithms employ the feature extraction phase, which is not required in CNNs.

The different imaging modalities such as magnetic resonance imaging (MRI), functional MRI (fMRI), X-ray imaging and computer tomography

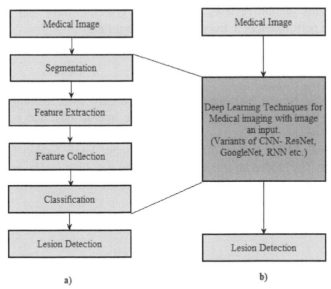

Medical image analysis a) before and b) after the introduction of Deep Learning

Figure 1: Deep learning with image input.

(CT), are essential with specific characteristics and finds its applications in the diagnosis of different human body structures. The medical image analysis involves diversified tasks like classification, recognition, segmentation, registration, enhancement, etc. Deep Neural Networks (DNN) are acquiring a great attention in analyzing the various imaging modalities as well as different phases of clinical diagnosis.

In the survey of a variety of anatomical structures across distinctive medical imaging modalities, deep learning techniques are preeminent with enhanced performance. This chapter discuss the medical image analysis with an emphasis on Convolutional neural networks (CNN) which can process the 2-dimensional as well as 3-dimensional data while capture the crucial image features that assists in reducing the number of parameters to be processed in learning mechanism.

Convolutional neural network architectures have exhibited cutting edge performances for medical image analysis and automated medical image diagnosis.

2 Network Structures

The Network Structure section introduces the major, popular network structures used for various image analysis tasks, their advantages, and shortcomings. This section reviews different CNN architectures that achieved remarkable performances in the medical imaging field.

2.1 Convolutional Neural Networks

Convolutional Neural Networks (CNN) are widely used learning models with three features: parameter sharing, sparse interactions, and equivalent representations.

A typical CNN consists of three layers-input layer, hidden layer and output layer. The hidden layer is composed of the convolutional layer, pooling and fully-connected layers. The architecture of CNN is described in Fig. 2.

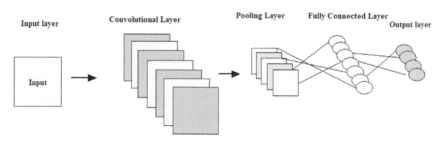

Figure 2: CNN architecture.

The high-level features from the input image are extracted by the filters employed in the convolutional layer. The aim of the convolution layer is to learn the weights of convolutional filters or kernels which perform the convolution operation on images. The CNN model converts the image into a sequence with the help of the convolution operation as,

$$C_j^l = f(\Sigma_{i \in M_j} C^{l-1} * W_{ij}^l + b_j^l) \qquad (1)$$

The convolution operation involves moving a small window over the image. For each step, the elements of the window and the corresponding elements of the image are multiplied and added up to result in a scalar value. The resultant of the convolution operation is another 2D grid, known as the activation map and sometimes also termed as the feature map.

2.1.1 Pooling Layer

The pooling layer is one of the primary building blocks in convolutional neural networks. It involves the down-sampling operation that predominantly leads to the reduction in dimensionality of the training data and also drastically reduces the number of parameters in the input. In addition to the aforementioned tasks, the pooling layer is associated with diminishing the height and width of the feature maps in CNN.

Usually two types of pooling strategies are employed in CNNs. Pooling layers operate on each feature map to yield a new feature map with enhanced features.

Max pooling: The most pronounced features of earlier feature maps are extracted by the pooling layer and yields a new feature map. The maximum element from the filter region is selected by the max pooling layer.

Average pooling: It enumerates the average value of all the elements present in the filter region. Averaging pooling generates less significant features compared with max pooling.

In addition to this global pooling is another layer present in a few of the deep networks. The pooling layer is not involved in learning parameters, hence in most of the networks it can be a part of convolutional layers.

2.1.2 Fully-Connected (FC) Layer

The fully connected layer is composed of the layers in which every input of the current layer is connected to every activation unit in the next layer. The fully connected layer generates a consolidated form of the input images. The primary objective of fully connected layers is performing the classification of features that are extricated from the input layers by using filters. Usually fully connected layers use 'soft-max' rule as the activation function that classifies the input based on the probability values ranging from zero to one. The frequently used activation functions in FC layers are sigmoid, tanh and parameterized ReLu.

In addition to the above mentioned layers, another layer called Batch Normalization (BN) is employed in recent CNN architectures. The BN layer is introduced to make the whole model more efficient by normalizing the activations to have a zero mean and unit variance.

The sophisticated CNN architectures composed of a convolutional layer inculcate with a stride of one that replaces the pooling layer to achieve an enhanced performance.

2.2 Deep Residual Network (ResNet)

A deep residual neural network (Balduzzi et al., 2017) is a specific form of CNN that adopts residual learning by stacking the building blocks along with a residual function, $f(x_k, \{W_k\})$ to be learned, of the form

$$y_k = f(x_k, \{W_k\}) + h(x_k) \qquad (2)$$

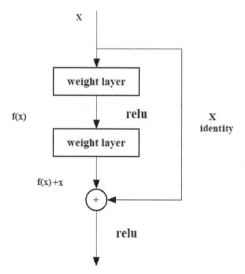

Figure 3: Residual block.

where, x and y are the input and output of *layer* k and $\mathbf{h}(x_k)$ is a linear projection that maps the residual function f and the input x.

Identity mapping is the core principle of Res-Net, which implies that the input of one layer is directly passed to the next layer. This is also termed as skip connection. The residual network uses a technique namely, skip connection in which the training process is skipped from a few layers and connects directly to the output as shown in Fig. 3. The advantage of adding a skip connection is, if any layer affects the performance of the network then that layer can be skipped by an appropriate regularization.

The commonly used architectures of ResNet varies with 50-layer, 101-layer, and 152-layer depths.

2.3 GoogleNet

GoogleNet, is a well-established CNN architecture and is prominent for its accuracy level that matches human performance levels. The GoogleNet architecture achieved first position in the ILSVRC2014 competition with an error rate of 6.67%. The essential components like pooling and convolutional layers in GoogleNet are similar to those of other CNN architectures. Christian Szegedy (szegedy2015going) proposed the concept of inception layers with different kernel sizes. The assimilation of 'Inception Layers' has enhanced the recognition accuracy of GoogleNet over the conventional architectures.

GoogleNet is composed of 22 layers, which was far greater than the other networks. Inception layers consist of variable receptive fields, employed to acquire the sparse correlation patterns from the feature maps. Andrej Karpathy designed a network with 22 layers depth but managed to reduce the number of parameters to four million when compared to AlexNet (60 million).

2.4 Fully-connected Convolutional Neural Networks (fCNN)

Recently, a prominent variation of CNN has become very popular. Fully convolution CNNs, have every node in one layer connected to every node in other layers. The fCNN makes no assumption about the input, and is less attractive compared with traditional CNNs for feature extraction.

For specific applications like segmentation to attain high-quality performances, a fCNN architecture is developed by replacing the fully connected layers in traditional CNNs with convolutional layers (Long et al., 2015). A fCNN with skip architecture combined with a rough output map is designed to extract the spatial information from shallower layers. Whereas, this structure was adopted from the basic CNN constructions like AlexNet and VGG-16 (Simonyan and Zisserman, 2014). For example, fFCNN-16S is constructed by up-sampling 16, fusion of output from 2 x up-sampled pool5 and pool4 as shown in Fig. 4.

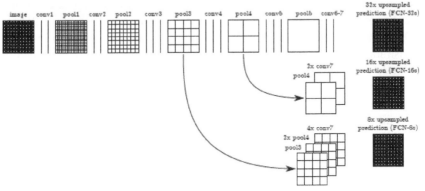

Figure 4: Skip architecture.

2.5 Applications of CNN in Medical Image Analysis

This section analyzes the application and contribution of CNNs for medical image analysis that includes various tasks like registration, segmentation, detection, and classification (Ker et al., 2017; Singha et al., 2021).

2.6 Registration

Image registration copes with the spatial alignment of two images, namely source and target to a common anatomical space by virtue of transformations. The source image can be a pre-operative MRI brain scan and the target image is the intra-operative version of the same, used to decide whether a remnant tumor is present. Registration methods find application in medical imaging in order to fuse the different imaging modalities such as, MRI and CT, and X-ray and CT.

Medical image registration methods can be discussed on the basis of different criteria like similarity index, dimensionality, time-series and registration. Specifically, for unimodal image registration such as CT-CT and MR-MR, similarity metric based methods have gained popularity over the last few years. The metrics usually employed are correlation coefficient (CC), mutual information (MI) and mean-square-distance (MSD). El-Zahraa et al. (El-Gamal et al., 2016) presented the various applications of medical image registration. A binary classifier is trained based on a deep similarity measure as proposed by Cheng et al. (Cheng et al., 2018). The correspondence between the CT-ME image patches is learned from the training process of the network. The coherence of the image patches is measured with a similarity index namely, the continuous probabilistic value.

The registration methods based on dimensionality criteria can be classified into spatial and time-series dimensions. Spatial dimensionality refers to the number of geometric dimensions of the image space which includes, 2D-to-2D, 3D-to-3D and 2D-to-3D. An overview of recent spatial dimension registration methods with specified regions of interest (ROI) is described in Table 1.

Time series dimension based registration methods depend on the alignment of medical images from the same or different modalities acquired over different time instances. This is quite useful in the evaluation of disease progress, particularly in assessing the treatment response. The methods

Table 1: Influential papers on image registration.

Reference	ROI	Dimension
(Sedghi et al., 2018)	Brain	2D-2D
(Ma et al., 2017)	Chest, Abdomen	2D-2D
(Kaiser et al., 2019)	Prostate	3D-3D
(Krebs et al., 2017)	Prostate	3D-3D
(Zheng et al., 2018)	Spine	3D-2D
(Miao et al., 2016)	Knee, hand	3D-2D

under registration basis criteria are grouped into extrinsic and intrinsic methods. Extrinsic methods deal with clearly visible artificial objects that are attached to the patient with the necessity to be accurately detectable in all the acquired modalities. Intrinsic methods deal with the knowledge acquired from plausible salient landmarks, binary divided structures or voxel image intensities provided by the patient.

2.7 Segmentation

Image segmentation is a trivial application of CNN architectures for medical image analysis. The segmentation performance in medical images has improved gradually over the conventional methods after introducing CNNs. Medical image segmentation includes different approaches based on threshold, region and clustering based techniques. The advancement in deep learning evidence for the great success in medical image segmentation with diverse neural network models. Segmentation is essential in crucial tasks pertaining to medical image analysis such as texture extraction in PET/SPECT images, automatic detection of vascular diseases and in radiology applications.

An end-to-end fully convolutional neural network (fCNN) achieved a remarkable performance improvement in segmentation. Since it can be trained pixels-to-pixels by taking an input of arbitrary size and produces an output with efficient inference and learning. In the recent past, deep learning methods have been proposed regularly for improving segmentation performances. A fCNN based architecture for medical image segmentation, with successive convolutional and pooling layers in order to reduce the dimensionality was proposed by Long et al. (Long et al., 2015). CNN architectures have been exploited for image intensity normalization in breast segmentation tasks. The automated segmentation techniques based on fCNN for breast density estimation and percent density (PD) estimation in mammograms have outperformed the state-of-the-art methods. Retinal blood segmentation in diabetic retinopathy is another challenging task where the CNN based architectures showed improved performance. In the early diagnosis of lung cancer, exact lung nodule segmentation is a crucial phase. To accurately estimate the nodule size in CT images and the segmentation contour, fCNN based techniques have shown a good performance. The tenfold cross validation with relatively more pooling layers (Zhao et al., 2019) and 3D fCNN (Sudre et al., 2017) are few related structures that improve the segmentation accuracy.

The segmentation of different anatomical structures like tumors, biological cells, lungs and membranes have been studied (Pereira et al., 2016; Akkus et al., 2017). Performing multitasks with segmentation and classification, regression or registration has the synergy to show a more precise segmentation performance.

2.8 Classification

In the medical image analysis framework, classification plays an important role for computer aided diagnosis (CAD). The classification process deals with assigning one or more labels to the object conforming to the features obtained from the medical image. The classification task involves an extensive span of applications ranging from determining the presence/absence of a disease to diagnosing the category of malignancy.

Diagnosing breast cancer, automated tissue characterization for early detection of lung infections, diabetic retinopathy and extraction of crucial features from functional MRI (fMRI) are quite a few to name out of a relatively large number of applications of CNN architectures. The recent developments in the diagnosis of common human anatomical regions such as brain, eyes and lungs are listed as follows.

2.8.1 Brain

Autism Spectrum disorder (ASD) in fMRI, classification of Alzheimer's disease (AD) and Mild-Cognitive Impairment (MCI) are the important tasks related to the brain activity that can be analyzed using CT scans or MRIs. CNNs have gained great importance in medical image analysis particularly related to classification tasks. For the assessment of the disease status related to the brain, 3D CNNs have achieved considerable accuracy (Mohammad jafari et al., 2016). Deep Belief Networks are quite useful to extract features from functional fMRI (fMRI) images, and MRI scans of patients with Huntington Disease and Schizophrenia. Even the 2D CNNs still work for analysis of 3D images by employing slice-by-slice operations. The accuracy of 2D networks is low compared to 3D networks, though the application of 2D structures reduces the computational complexity.

2.8.2 Eye

The digital photographs of the fundus eye are employed to diagnose Diabetic retinopathy (DR) using CNNs. Islam et al. (Islam et al., 2018) proposed a CNN model for early detection of DR in retinal fundus images. Pratt et al.

(Pratt et al.) trained a CNN with 10 convolutional layers and obtained 75% accuracy in classifying DR into five clinically used categories. AlexNet, VGG-19 and ResNet50 are the other structures that outperform the conventional methods.

2.8.3 Lungs

Shen et al. (Shen et al., 2015) used 1010 CT scans that are labelled. The Lung Image Database Consortium (LIDC-IDRI) is employed to divide the lung nodules into benign or malignant. The methodology involves classification based on Support Vector Machine (SVM) and Random Forest (RF) classifiers combined with CNN. They used 3 parallel CNNs with 2 convolution layers each, with each CNN taking image patches at different scales to extract features.

In addition, computer-aided diagnosis of liver cancer, gallstone, prostate cancer, uterine fibroid, and bladder cancer is performed with the aid of CNN (Sasikumar and Rajendran, 2018; Cha et al., 2017).

2.9 Detection

The role of detection in medical image analysis includes the extraction of the region of interest (ROI) or detection of instances of lesions. Apart from classification, detection tasks extract the most valuable information and locate the lesions from medical images of human organs namely, eyes, lungs, breast, stomach, bladder, prostate and kidney.

The list of representative works for detection using CNNs are presented in Table 2.

Table 2: Representative works on detection.

Author	Method	Image modality
(Gu et al., 2018)	3D CNN	CT (lung)
(Sumathipala et al., 2018)	Deep CNN	MRI (Prostate)
(Togo et al., 2019)	Deep CNN	X-ray
(Cireşan et al., 2013)	CNN	Breast Histology Image
(Sakai et al., 2018)	CNN	Endoscopy (Gastric)
(Schlegl et al., 2018)	CNN	OCT (eye)
(Islam and Zhang, 2018)	CNN	MRI (Alzheimer's Disease (AD))
(Chiang et al., 2018)	3D CNN	ABUS* (Breast cancer)

* Automated breast ultra-sound.

2.10 Data Sets

A massive number of medical image training data sets are necessary to train the weights and various parameters of a convolutional neural network and its derived architectures. The challenge of a lack of datasets is addressed to some extent by providing publicly available sources as well as open-access medical image training sets.

Links for some of the publicly available data sets:

MURA - Musculo skeletal radiographs-
https://arxiv.org/abs/1712.06957NIH-X-ray-
https://www.kaggle.com/nih-chest-xrays/data
NCI NLST - https://biometry.nci.nih.gov/cdas/dataset/nlst
TCGA-GBM - https://wiki.cancerimagingarchive.net/display/public/TCGA-GBM

Table 3, presents a list of some open-access data sets that are frequently used in training CNNs.

Table 3: Medical image datasets (Open-Access).

Data set	Remarks
FITBIR	Federal interagency traumatic brain injury research
ADNI	Alzheimers disease neuroimaging initiative
BraTS2015	Brain tumor segmentation
LIDC/IDRI	Lung nodule detection
MIDAS	Medical and non-medical image data
SCR	Segmentation in chest radiographs
OASIS	MRI data
MITOS	Breast histology images
PROMISE2012	MRI prostate images
TCIA	Cancer imaging archive
DIARETDB1	Retinal Images for diabetic retinopathy
Stanford AI	Mix of CheXpert

2.11 Challenges & Future Directions

The major challenges in the application of CNNS for medical image diagnosis and analysis are discussed in this section. The common challenge is selecting a best set of hyper parameters through optimization, learning rate, batch size and the learning models. The second challenge is a lack of labelled medical image training data sets. Another problem related to the medical images data is class imbalance. The data imbalance issue is related to the number of images being skewed across the available classes. Training efficiency is another challenge in CNN based medical image analysis. Improving the learning efficiency is crucial since the CNN consists of a huge number of parameters, another significant area of research. An issue related to the training data is the unavailability of an adequate number of labels for the medical images. However, very few researchers are working in this area, adding labels is a laborious and stagnant task. Dimensionality of the medical images is another crucial problem related to the performance of CNNs. There is a necessity to design the higher order deep learning structures to achieve the desirable performance with high-dimensional data.

The lack of training data problem can be alleviated by adopting the transfer learning mechanism in CNNs. Investigating the optimal set of regularization terms for transfer learning is an interesting research area in medical imaging. Data augmentation is an impressive mechanism to deal with the problem of data imbalance but at the cost of overfitting. Data augmentation refer to the production of more training images of uncommon data. Tensor-based training models can be explored to deal with the multi modal and high dimensional medical data.

3 Conclusion

Convolutional neural networks suggest their potential in medical imaging with promising results. This chapter reviews the constructional features of convolutional neural networks and its variants that are suitable for medical image diagnosis. Further, convolutional neural networks have been successfully utilized for medical image registration, segmentation, detection and classification. Even though the earlier works yield promising results there is still a considerable number of issues to be resolved in deep learning methods concerned with the medical image field. The challenges connected with the medical image applications are discussed. Finally, the possible future research directions with an objective of addressing diversified issues are presented.

References

Alireza Sedghi, Jie Luo, Alireza Mehrtash, Steve Pieper, Clare M. Tempany et al. (2018). Semi-supervised deep metrics for image registration. *arXiv preprint arXiv:1804.01565*.

Carole H. Sudre, Wenqi Li, Tom Vercauteren, Sebastien Ourselin and Jorge M. Cardoso et al. (2017). Generalised dice overlap as a deep learning loss function for highly unbalanced segmentations. In *Deep Learning in Medical Image Analysis and Multimodal Learning for Clinical Decision Support*, pp. 240–248. Springer.

Dan C. Cireşan, Alessandro Giusti, Luca M. Gambardella and Jürgen Schmid-huber. (2013). Mitosis detection in breast cancer histology images with deep neural networks. In *International Conference on Medical Image Computing and Computer-Assisted Intervention*, pp. 411–418. Springer.

David Balduzzi, Brian McWilliams and Tony Butler-Yeoman. (2017). Neural taylor approximations: Convergence and exploration in rectifier networks. In *International Conference on Machine Learning*, pp. 351–360. PMLR.

Fatma El-Zahraa, Ahmed El-Gamal, Mohammed Elmogy and Ahmed Atwan. (2016). Current trends in medical image registration and fusion. *Egyptian Informatics Journal*, 17(1): 99–124.

Grant Haskins, Jochen Kruecker, Uwe Kruger, Sheng Xu, Peter A. Pinto et al. (2019). Learning deep similarity metric for 3d mr–trus image registration. *International Journal of Computer Assisted Radiology and Surgery*, 14(3): 417–425.

Harry Pratt, Frans Coenen, Deborah M. Broadbent, Simon P. Harding and Yalin Zheng et al. (2016). Convolutional neural networks for diabetic retinopathy. *Procedia Computer Science*, 90: 200–205. ISSN 1877-0509. doi: https:// doi.org/10.1016/j.procs.2016.07.014. URL https:// www.sciencedirect. com/science/article/pii/S1877050916311929. 20th Conference on Medical Image Understanding and Analysis (MIUA 2016).

Jiannan Zheng, Shun Miao, Jane Z. Wang and Rui Liao. (2018). Pairwise domain adaptation module for cnn-based 2-d/3-d registration. *Journal of Medical Imaging*, 5(2): 021204.

Jonathan Long, Evan Shelhamer and Trevor Darrell. (2015). Fully convolutional networks for semantic segmentation. In *Proceedings of the IEEE Conference on Computer Vision and Pattern Recognition*, pp. 3431–3440.

Julian Krebs, Tommaso Mansi, Hervé Delingette, Li Zhang, Florin C. Ghesu et al. (2017). Robust non-rigid registration through agent-based action learning. In *International Conference on Medical Image Computing and Computer-Assisted Intervention*, pp. 344–352. Springer.

Justin Ker, Lipo Wang, Jai Rao and Tchoyoson Lim. (2017). Deep learning applications in medical image analysis. *IEEE Access*, 6: 9375–9389.

Jyoti Islam and Yanqing Zhang. (2018). Early diagnosis of alzheimer's disease: A neuroimaging study with deep learning architectures. In *Proceedings of the IEEE Conference on Computer Vision and Pattern Recognition Workshops*, pp. 1881–1883.

Kai Ma, Jiangping Wang, Vivek Singh, Birgi Tamersoy, Yao-Jen Chang et al. (2017). Multimodal image registration with deep context reinforcement learning. In *International Conference on Medical Image Computing and Computer-Assisted Intervention*, pp. 240–248. Springer.

Kaiser, A., Haskins, C., Siddiqui, M. M., Hussain, A. and D'Adamo, C. (2019). The evolving role of diet in prostate cancer risk and progression. *Current Opinion in Oncology*, 31(3): 222.

Kenny H. Cha, Lubomir Hadjiiski, Heang-Ping Chan, Alon Z. Weizer, Aj- jai Alva et al. (2017). Bladder cancer treatment response assessment in CT using radiomics with deep-learning. *Scientific Reports*, 7(1): 1–12.

Ren Togo, Nobutake Yamamichi, Katsuhiro Mabe, Yu Takahashi, Chihiro Takeuchi et al. (2019). Detection of gastritis by a deep convolutional neural network from double-contrast upper gastrointestinal barium x-ray radiography. *Journal of Gastroenterology*, 54(4): 321–329.

Sanaz Mohammadjafari, Mucahit Cevik, Mathusan Thanabalasingam and Ayse Basar. Using protopnet for interpretable alzheimer's disease classification.

Sasikumar, D. and Rajendran, P. (2018). Identification of uterine fibroids using enhanced image mining techniques: Bio-inspired xenogenetic based extreme learning neural network classification with improved fireflies hausdorff distance. *Current Medical Imaging*, 14(5): 822–830.

Sérgio Pereira, Adriano Pinto, Victor Alves and Carlos A. Silva. (2016). Brain tumor segmentation using convolutional neural networks in MRI images. *IEEE Transactions on Medical Imaging*, 35(5): 1240–1251.

Sheikh Muhammad Saiful Islam, Md Mahedi Hasan and Sohaib Abdullah. (2018). Deep learning based early detection and grading of diabetic retinopathy using retinal fundus images. *arXiv preprint arXiv:1812.10595*.

Shun Miao, Z. Jane Wang and Rui Liao. (2016). A cnn regression approach for real-time 2d/3d registration. *IEEE Transactions on Medical Imaging*, 35(5): 1352–1363.

Simonyan, K. and Zisserman, A. (2014). Very deep convolutional networks for large-scale image recognition. *arXiv preprint arXiv:1409.1556*.

Singha, S., Pasupuleti, S., Singha, S. S., Singh, R. and Kumar S. (2021). Prediction of groundwater quality using efficient machine learning technique. *Chemosphere*, 276: 130265

Szegedy, C., Liu, W., Jia, Y., Sermanet, P., Reed, S. et al. (2015). Going deeper with convolutions. In *Proceedings of the IEEE Conference on Computer Vision and Pattern Recognition*, pp. 1–9.

Thomas Schlegl, Sebastian M. Waldstein, Hrvoje Bogunovic, Franz Endstraßer, Amir Sadeghipour et al. (2018). Fully automated detection and quantification of macular fluid in oct using deep learning. *Ophthalmology*, 125(4): 549–558.

Tsung-Chen Chiang, Yao-Sian Huang, Rong-Tai Chen, Chiun-Sheng Huang, Ruey-Feng Chang et al. (2018). Tumor detection in automated breast ultrasound using 3-d cnn and prioritized candidate aggregation. *IEEE Transactions on Medical Imaging*, 38(1): 240–249.

Wei Shen, Mu Zhou, Feng Yang, Caiyun Yang and Jie Tian et al. (2015). Multi-scale convolutional neural networks for lung nodule classification. In *International Conference on Information Processing in Medical Imaging*, pp. 588–599. Springer.

Xinzhuo Zhao, Wenqing Sun, Wei Qian, Shouliang Qi, Jianjun Sun et al. (2019). Fine-grained lung nodule segmentation with pyramid deconvolutional neural network. In *Medical Imaging 2019: Computer-Aided Diagnosis*, volume 10950, page 109503S. International Society for Optics and Photonics.

Yohan Sumathipala, Nathan S. Lay, Baris Turkbey, Clayton Smith, Peter L. Choyke et al. (2018). Prostate cancer detection from multi-institution multiparametric mris using deep convolutional neural networks. *Journal of Medical Imaging*, 5(4): 044507.

Yoshimasa Sakai, Satoko Takemoto, Keisuke Hori, Masaomi Nishimura, Hiroaki Ikematsu et al. (2018). Automatic detection of early gastric cancer in endoscopic images using a transferring convolutional neural network. In *2018 40th Annual International Conference of the IEEE Engineering in Medicine and Biology Society (EMBC)*, pp. 4138–4141. IEEE.

Yu Gu, Xiaoqi Lu, Lidong Yang, Baohua Zhang, Dahua Yu et al. (2018). Automatic lung nodule detection using a 3d deep convolutional neural network combined with a multi-scale prediction strategy in chest cts. *Computers in Biology and Medicine*, 103: 220–231.

Zeynettin Akkus, Alfiia Galimzianova, Assaf Hoogi, Daniel L. Rubin, Bradley J. Erickson et al. (2017). Deep learning for brain mri segmentation: State of the art and future directions. *Journal of Digital Imaging*, 30(4): 449–459.

Chapter 5

Ulcer and Red Lesion Detection in Wireless Capsule Endoscopy Images using CNN

Said Charfi,[1,*] *Mohamed El Ansari,*[2] *Ayoub Ellahyani*[3] and *Ilyas El Jaafari*[3]

ABSTRACT

Traditional endoscopies have dominated the field of the gastro-intestinal tract investigation for decades. Nevertheless, since its emergence, the wireless capsule endoscopy imaging technique has gained more popularity as it can visualise the entire gastro-intestinal tract including the small bowel. However, this technique produces immense amount of images which have to be reviewed

---

[1] LabSIV, Department of Computer Science, Faculty of Science, Ibn Zohr University, BP 8106, 80000 Agadir, Morocco.
[2] Informatics and Applications Laboratory, Department of Computer Science, Faculty of Science, My Ismail University Meknès, Morocco LabSIV, Department of Computer Science, Faculty of Science, Ibn Zohr University, BP 8106, 80000 Agadir, Morocco.
[3] LabSIE, Department of Mathematics and Computer Science, multidisciplinary faculty, Ibn Zohr University, BP 638, 45000 Ouarzazate, Morocco.
* Corresponding author: charfisaid@gmail.com

by the physicians. This task is exhaustive and time consuming. This chapter presents a novel approach adopted for colon red lesion and ulcer abnormalities detection using images issued by WCE. A simple Convolutional Neural Network architecture is proposed. Besides, we incorporated Parametric Rectified Nonlinear Unit (PRenu) activation function. Extensive experiments have been conducted on two datasets in order to show the efficiency of the proposed method.

1 Introduction

Wireless capsule endoscopy (WCE), a gastrointestinal visualization technique, has become more reliable compared to traditional endoscopies. It can access and view the entire small sized bowel in the human body with less invasiveness and pain. However, the process of investigating the meticulously produced endoscopy videos, from the gastrointestinal examinations, is tiring and time consuming. This chapter addresses the problem of colon abnormalities detection and identification from images produced after a capsule endoscopy. It details the approach followed for this purpose. In the present method, a simple CNN model for red lesion and ulcer disease discrimination in WCE images is presented. The model is strengthened by the new Parametric Recitified Non linear Unit (PRenu) activation function. The presented scheme has been assessed with experimentations in several medical databases.

Since its innovation in 2001, many methods for automatic abnormalities detection in WCE images have been proposed. Most of these approaches are based on extracting the features describing the abnormality from WCE images. Despite, several state-of-the-art methods claim to have reached a very good detection accuracy, the absence of any standard dataset in the WCE field makes their performance subjective. Therefore, researchers are still working on this issue. In this chapter, we begin by defining the role of Computer-Aided Diagnosis (CAD) systems and we present an overview of different GI tract visualization techniques. This introduction is also devoted to related works on the problem of detecting abnormalities from WCE images. Firstly, a definition of the capsule endoscopy technique is given. Secondly, we have stated the most relevant works focusing on ulcer diseases. Afterwards, red lesion disease is introduced. Finally, methods in the literature based on deep learning architectures are stated. The proposed approach is detailed in Section 2 followed by the experimental results in Section 3. Then, we end this chapter by a brief conclusion in Section 4.

1.1 Computer-aided Diagnosis Systems

According to the recently published Food and Drug Administration guidance (US Food, Drug Administration et al., 2012), computer-aided diagnosis systems can be divided in two categories. The first category includes, computer-aided diagnosis (CADx) systems aiming at analyzing a radiographic output in order to evaluate the possibility that the attribute requires a disease process (e.g., benign versus malignant), or intend to specify disease type (i.e., specific diagnosis or differential diagnosis), severity, stage, or recommended intervention. For this purpose a conventional CADx is often designed in three phases: feature extraction, feature selection, and classification. The first phase is often viewed as the most important phase as it probably facilitates the other phases. Feature extraction's main objective is to extract discriminative features depending on each specific problem. Briefly speaking, these features can be divided into textural and morphological ones. However, the engineering of robust features is influenced by the latter phases of feature selection and attribute classification. In addition, the feature extraction phase is mostly problem-oriented. This phase is important to achieve accurate normal/abnormal differentiation. In practice, these three steps are investigated separately and then combined together for the overall CADx performance adjustment. The second category called computer-aided detection (CADe) which is defined as a technology aiming to design helpful systems to decrease observational oversights. It especially reduces the false negative rates when clinicians review medical images. In addition, it intends to identify, mark, highlight, or in any other manner direct attention to portions of an image that may reveal specific abnormalities during the interpretation of patient radiology images. In the past decades, computers had been increasingly used to help radiologists in the acquisition, storage and management of medical images (e.g., CT, MRI, computed radiography). More recently, computer programs have been developed detecting abnormalities, thus, approved for use in clinical practice. In order to evaluate the CADe system, firstly the radiologist analyses the test results, then, compares them with results obtained by the CADe system. Finally, an evaluation report is made. Besides, post the CAD system design, its performance can be assessed by many other methods. These methods for CAD evaluation includes the in laboratory analysis of data produced by CAD system, or by setting a test procedure, or by evaluating the contribution of the CAD system to the radiologist's production in a real clinical environment (Castellino, 2005; Chang, 2015). The methods mentioned in this chapter therefore include both CADe and CADx (CAD).

1.2 Capsule Endoscopy

Virtual endoscopy techniques such as X-ray, MRI and CT are probably the most exploited visualization techniques in the diagnostic procedure in GI medicine. The procedure followed by these endoscopy techniques is uncomfortable for the patient and provides limited views. Besides, it exposes the patient to a possible risk of lesions because of the endoscope. Recently, endoscopy techniques have been enriched by a promising new method, capable of giving a detailed visualization and study of mucosal layers. In addition, the small intestine can now be examined by this technology called capsule endoscopy. The procedure is very simple and includes a device of small size. The patient undergoing the procedure swallows the capsule which traverses the gastrointestinal tract. The device visualizes and captures images of the gastrointestinal tract, small intestine included, and at the end a video is obtained.

1.3 Computerized Methods in Capsule Endoscopy

In this section, we give a literature survey of some computerized methods used for colon abnormalities detection. Particularly, we focus on those using medical images acquired by wireless capsules.

1.3.1 Ulcer Detection Approaches

The digestive tract is lined by a mucus membrane. Ulcers are sores that break out on the mucus membrane of the digestive tract. Figure 1 shows a WCE image with ulcer disease. Automatic ulcer detection has been excessively investigated in the literature. In (Li and Meng, 2009), textural features extracted using local binary patterns after multiresolution analysis using curvelet has been proposed. Similarly, the authors have presented a wavelet based local binary pattern scheme to discriminate ulcer regions from normal ones in patches selected from CE images. In (Charisis et al., 2012), authors investigated common structural components of WCE frames using Bidimensional Ensemble

Figure 1: WCE image with ulcer.

Empirical Mode Decomposition (BEEMD). Based on the results of the aforementioned investigation, authors try to reconstruct a new refined image. Thereafter, the intrinsic second/higher-order correlation of the original frames and refined ones, is exploited for textural features extraction using the lacunarity index. In another work (Charisis et al., 2012), the same authors extended their previous work. In the latter one, CE images were analyzed, in RGB, HSV and CIE Lab colour spaces, for textural-colour characteristics extraction. Thus, the distribution of structural information of healthy and abnormal regions on RGB, HSV and CIE lab colour spaces has been studied through the investigation of colour texture features. Besides, textural characteristics were extracted, from normal and ulcerous regions in WCE images, using differential lacunarity analysis. Again, BEEMD was used as pre-processing phase. In (Lecheng et al., 2012), a combination of local features with the feature fusion technique through the bag-of-words method, has been the basis of a new image processing method. Eid et al. (2013) suggested deriving textural features using the discrete curvelet transform. Afterwards, they extracted textural information by computing the lacunarity index for each discrete curvelet subband. In (Szczypinski et al., 2014), single pixels were classified based on the neighbors' textural information. Then, the authors suggested segmenting images into homogeneous regions. In another work, WCE images were rotated by a colour rotation operation as a preprocessing step. Then, a uniform rotation invariant local binary pattern operator was utilized for feature extraction (Charisis et al., 2013). Log Gabor filter and Contourlet transform were utilized to discern images showing ulcer abnormalities from the ones without any abnormality (Koshy and Gopi, 2015). Aimed at detecting several diseases in endoscopy frames, a local binary pattern histogram representing a set of image textons and a combination of Leung and Malik filter bank (set of image filters) were analyzed in (Nawarathna et al., 2014) for discriminative features extraction. Colour and textural features were extracted in (Yeh et al., 2014) in order to detect ulcers in WCE images, hence, decide on the small intestine status. An automated detection of lesions in CE frames was published in (Yuan et al., 2015). A masking approach with a SVM classifier in the RGB colour plane were presented in (Suman et al., 2016). In (Ghosh, 2016), black portions of CE images are removed, then a combination of RGB colour plane histograms were applied for feature extraction. WCE images texture information was exploited in the (Charfi et al., 2018). In this work, authors suggested the use of a LBPV descriptor based on the discrete wavelet transform for multi-resolution analysis for the detection of several abnormalities in WCE frames. The same authors, proposed a new feature extraction approach to distinguish inflammation and ulcer regions in CE frames (Charfi and Ansari, 2020). This approach was also exploited for polyp detection in images issued by a colonoscopy procedure. Recently, a gray scale was used to compute the histograms from pixel values in (Kundu et al., 2016) instead of the RGB colour space. In (Kundu and Fattah, 2017), authors suggested the use of asymmetric RGB indexed images, exploited for colour histogram generation as a feature descriptor instead of using the original histogram. Hence, priority will be given to the colour channel carrying more information. In order to detect ulcerous frames, authors in (Charfi and Ansari, 2017) suggested extracting textural information using completed local binary patterns and derived colour features with the application of global local oriented edge magnitude patterns. A scheme based on the Laplacian pyramid along with completed local binary patterns was exploited for WCE frames multi-scale analysis was stated in (Souaidi and Ansari, 2019). The Y plane of the YIQ colour space was investigated in (Kundu et al., 2017) for ulcer identification. A new approach based on segmentation of ulcerous regions using saliency maps and feature extraction using a combination of state-of-the-art descriptors can be found in (Charfi et al., 2019). A multi-scale analysis-based grey-level co-occurrence matrix was proposed in (Souaidi and Ansari, 2019). Table 1 depicts a comparison of a few approaches proposed for ulcer detection from WCE images. In addition, a comparison of methods aiming to detect various abnormalities is depicted in Table 2.

Table 1: Comparison of ulcer detection methods (%).

Method	Ulcerous images	Normal images	Accuracy	Sensitivity	Specificity
Szczypiński et al., 2014	110	110	87.27	88.64	85.75
Koshy and Gopi, 2015	65	72	94.16	96.92	91.67
Kundu et al., 2016	75	75	87.23	85.13	90.42
kundu et al., 2017	75	75	87.49	83.68	91.08
Kundu and Fattah, 2017	75	75	87.09	84.08	90.93
Ghosh et al., 2016	260	650	92.01	89.90	91.32
Koshy and Gopi, 2015	36	500	-	97.00	66.00
Yeh et al., 2014	190	258	85.71	87.90	84.11
Yuan et al., 2015	170	170	92.65	94.12	91.18

Table 2: Comparison of methods aiming at various abnormalities detection (%).

Method	Abnormalities	# of abnormal frames	# of normal frames	Metrics
Maghsoudi et al., 2018	tumor	43		accuracy/94.61
	bleeding	33	44	sensitivity/96.71
	others	113		specificity/93.81
Nawarathna et al., 2014	erythema	125		
	blooding	125		recall/92.00
	polyp	125	1200	specificity/91.8
	ulcer + erosion	175		
Sindhu and Valsan, 2017	polyp	187		accuracy/97.50
	tumor	100	148	sensitivity/93.4
				specificity/98.8
Yuan et al., 2016	bleeding	500		
	polyp	500	500	accuracy/88.61
	ulcer	150		

1.3.2 Red Lesion

Red lesions can be defined as a large, heterogeneous group of disorders of the oral mucosa. They exhibit a red colour that can be due to many reasons including inflammation, dilatation of bloods vessels, thin epithelium or sometimes it occurs when the number of blood vessels increases or when the blood is extravased into the oral soft tissues. Figure 2 shows a WCE image with red lesions. Recently, Amiri et al. (2022) proposed a novel set of features for bleeding and angiodysplasia lesion detection in WCE frames. Their method can be split into two stages. In the first one, they extract the region of interest using the expectation maximization segmentation algorithm followed by some morphological operations applied to the winner segment. In the second stage, they derive two sets of features: a set of statistical features extracted from regions of interest from different components of various colour planes, i.e., 'a' in CIELab, 'b' in CIELab, 'Cr' in YCbCr and 'G' in RGB. Set 2 contains histogram-based features extracted from the 'a' component of the CIELab colour space. Then, the two sets of attributes are combined and fed to a multilayer perceptron classifier of normal/abnormal classification. Except for the work carried out by in Coelho et al., 2018 no other research, we have come across, had investigated the detection of red lesions in CE frames using deep learning models. The authors have employed a U-net deep learning model in order to segment the red lesion regions. The U-net model is a CNN variant.

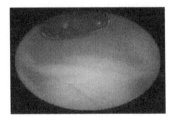

Figure 2: WCE image with red lesion.

1.3.3 Deep Learning based Approaches

In this section, we investigate the usage of the emerging deep learning algorithms and state a few works that adapted them in the WCE field. In a related work, Zou et al. (Zou et al., 2015) classified digestive organs using a deep convolutional neural network model. In (Yu et al., 2015), authors classified CE frames using hybrid CNN with an extreme learning machine. In the aforementioned work, a cascaded extreme learning machine was employed in a Deep CNN classification system replacing the well known fully connected layer. Generic characteristics were targeted in (Zhu et al., 2015) using the Deep CNN model for lesion identification. In this work, authors exploited the robustness of the SVM classifier instead of the fully connected layer. A Deep CNN model was applied for bleeding detection and motility characterization in the small intestine in both (Segu et al., 2016) and (Jia and Meng, 2016). In (Yu et al., 2012) authors proposed a novel deep feature learning method, named stacked sparse autoencoder with an image manifold constraint. A novel offline and online 3D deep learning integration framework by leveraging the 3D fully convolutional network (3D-FCN) was stated in (Yu et al., 2016). In (Sekuboyina et al., 2017), frames were split into many patches. Then, CNN was employed for the extraction of colour attributes from each patch. Aiming at automatic recognition of polyp disease from a WCE frame, an approach built upon CNN architecture was published in (Xu et al., 2016). In this paper, Alexnet, a variant of CNN, was deployed to discern images showing polyp abnormalities from the ones without any abnormality. A novel deep hookworm recognition framework has been stated in (Hajabdollahi et al., 2020) for capsule endoscopy images. He et al. suggested the modelling of tubular patterns and visual appearances of the aforementioned disease. Some interesting works reviewing the CE retention and quantitative analysis can be found in (Rondonotti, 2017; Ciaccio et al., 2017). Coelho et al. in (Coelho et al., 2018) proposed the use of a Unet model for red lesion segmentation in WCE frames. In (Pannu et al., 2020) the authors proposed a CNN architecture for hemorrhage detection in capsule endoscopy frames. They augmented the dataset using synthetic augmentation. Hajabdollahi et al. (Hajabdollahi et al., 2020), presented a two branch framework to derive the bifurcated structure. These two branches are independently trained for each disease, with one branch for segmentation and the other one for classification. Finally, the two branches are merged to form an overall system. A method using CNN as a backbone and making use of LSTM after each pooling layer has been presented in (Ozturk and Ozkaya, 2021) for disease classification in colonoscopic images. Then, it combines the LSTM layer features for classification. Guo et al. (Guo and Yuan, 2020) presented a method for CE frames classification in which they used Adaptive Aggregated Attention with a semi-supervised approach. In this scheme, they remove the circular boundaries and black background present in WCE frames using a deformation field for image preprocessing. Afterwards, they proposed a two stage feature extraction method. In the first stage, global dependencies and context information incorporated using Adaptive Aggregated Attention are employed. This stage, highlights the most meaningful parts in the image. In the second stage, the computed parts (regions) are exploited for robust and accurate abnormal frames classification. Finally, these two stages are jointly optimized. Authors proposed to minimize the the Jensen-Shannon divergence loss along with the discriminative angular loss, presented in the work, with labeled and unlabeled data. A method built upon fusion of geometric features and CNN was stated in (Sharif et al., 2021). Firstly a contrast-enhanced colour features approach was proposed for abnormal regions' segmentation in WCE images. Secondly, the authors derived geometric characteristics of the disease from segmented regions. Then, VGG19 and VGG16 deep CNN features fusion was applied based on the Euclidean Fisher Vector. Afterwards, feature selection is performed on features previously fused with geometric features using the conditional entropy method. Then, the best selected features are fed to the K-Nearest Neighbour classifier. In (Wang et al., 2020), a multi-scale context guided deep network is stated for lesion segmentation in colonoscopy videos. In this work, authors extract the global structure and high-level semantic information in one subnetwork (Gnet). Besides, they derive the multi-scale appearance information from the shallow layer of Gnet. In addition, semantic information is exploited from the deep layer of the Gnet. These subnets are combined together in a cascaded manner. In (Soffer et al., 2020), a review of deep learning applications in the field of WCE was carried out. In addition, surveys on deep learning methods for WCE image analysis were published in (Muruganantham and Balakrishnan, 2021; Rahim et al., 2020).

2 Proposed Method

In this section, we present methods devoted to colon abnormalities detection using WCE images. The scheme proposed serves to recognize red lesions and ulcers from WCE images. Based on a simple CNN architecture along with the PRenu activation function proposed in (El Jaafari et al., 2021), the presented scheme has been built. The methodology followed, in this work, is detailed below.

2.1 Preprocessing

Before passing the images to the CNN model, data augmentation was performed. The original WCE images were rotated by 40 degrees, shifted to the right and left, zoomed and distorted along the x axis by 0.2. Then the horizontal flip and filling, using the nearest pixel, are carried out. Finally, the images are converted to the HSV colour space. In order to perform all these image transformations, keras proposed functions are employed.

2.2 Convolutional Neural Network Architecture

Simple CNN architectures, having few layers, have demonstrated a better performance in many classification problems (Sekuboyina et al., 2021). Hence, our proposed CNN model consists of an input layer, output layer, and few hidden layers as intermediate layers (Fig. 3). The proposed model can be seen as a combination of three blocks, each block possesing two convolutional layers followed by a max pooling and drop out layers. Different from other works, the

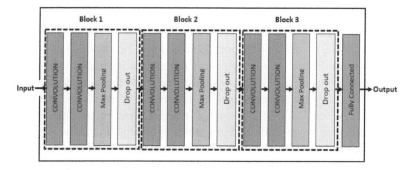

Figure 3: Scheme of the proposed CNN model.

convolutional layers employed in this model are followed by the new PRenu activation function. In addition, automatic feature extraction is performed and simplified by recognizing simple patterns and obscure ones using many filters. Then, max pooling is applied to pick-out the most prominent characteristics of the feature map output by the secondary convolutional layers. At the end of each block, a drop out layer is added to ensure model regularization.

2.3 Parametric Rectified Non Linear Unit

The Parametric rectified nonlinear unit (PRenu) activation function is nearly similar to Relu. The used activation function and its derivative are defined in the following Eqs. (1) and (2), respectively:

$$\phi(x) = \begin{cases} x - \alpha \log(x+1) & x > 0 \\ 0 & x \leq 0 \end{cases} \qquad (5.1)$$

$$\phi'(x) = \begin{cases} 1 - \alpha(x+1) & x > 0 \\ 0 & x \leq 0 \end{cases} \qquad (5.2)$$

PRenu returns $x - \alpha \log(x+1)$ for positive values and zero otherwise. As can be deduced from Eq. (2), PRenu is the same as the Relu activation function when α is set to 0. Different to Relu activation function, the PRenu activation function outputs more sparse representation in particular when more units that exist in a layer have x less than zero. Therefore, convolution neural networks learn more attributes as the sparsity and PRenu range varies between zero and infinity. Moreover, as the absolute value of x increases the gradient becomes higher, hence, the Prenu likelihood of the gradient to vanish is reduced. Different to Relu, all positive neurons are activated based on the neuron value, by the PRenu activation function, using a modified gradient. It also adjusts the chosen neuron weights based on their participation (El Jaafari et al., 2012).

3 Experimental Results

Aiming at evaluating the performance of the proposed scheme, we have tested it on two different colon diseases i.e., ulcer and red lesion, depicted in two datasets. The first one is a patches image of size 128×128 and composed of 1594 normal images and 4584 abnormal ones after data augmentation (Xu et al., 2018). The second one contains 4000 normal images and 6401 abnormal ones. The images had the size of 320×320 then resized to 128×128. It is available at https://rdm.inesctec.pt/dataset/nis-2018-003. Figures 4 and 5 show examples of the datasets used.

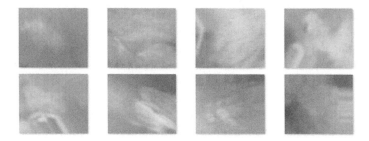

Figure 4: Examples of images containing ulcer disease from the first dataset.

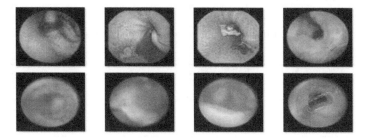

Figure 5: Examples of images containing red lesion disease from the second dataset.

3.1 Implementation Details

Two datasets, one containing ulcerous and the other one consisting of red lesion frames, were exploited for the proposed CNN model training. These databases were split into 80% for training and 20% for testing. Similarly the training dataset has been split into training and validation. 75% were used for training and 25% for validation. Binary cross entropy cost function was deployed in the training, Adam optimizer and 3 fold cross validation in 100 epochs for the first dataset and 50 for the second. Regarding the learning rate, the convolutional 1E-4 value was chosen. In addition, accuracy measure was used for model evaluation. The network was implemented in Python 3.6 and all experiments were performed on a machine provided by google Colab with GPU and 12GB RAM. The CNN was implemented using Keras with TensorFlow as backend. The number of parameters issued from the proposed CNN model is 164,514.

3.2 Evaluation on First Dataset

This section presents in detail the results of applying the proposed model. In particular, WCE frames exhibiting ulcerous regions and ones without any abnormality were utilized. These images are extracted from videos issued after a capsule endoscopy procedure. Table 3 shows the training and validation accuracies and losses obtained in the first dataset. Besides, the accuracy achieved in the testing dataset. The values are given for each fold. The proposed model achieved 97.16% and 91.04% average accuracy in the training and validation datasets, respectively. In terms of loss, the proposed approach achieved an average of 0.0793% and 0.2951% in the training and validation datasets, respectively. Besides, it reached 96.51% (0.1085) compared to 96.20% (0.1897), achieved by the Relu activation function, in accuracy (loss) metric. In addition, Fig. 6 exhibits the corresponding training and validation curves for the loss and accuracy metrics obtained in the first dataset. These curves depict the history of training and validation of the second fold.

Table 3: Performance of the proposed method on first dataset.

Fold	Training accuracy (loss)	Validation accuracy (loss)	Testing accuracy (loss)	Testing accuracy (loss) (Relu)
Fold 1	0.9664 (0.0819)	0.9185(0.2489)		
Fold 2	0.9759 (0.0737)	0.8879 (0.4135)	0.9651 (0.1085)	0.9620 (0.1879)
Fold 3	0.9725 (0.0823)	0.9250 (0.2229)		

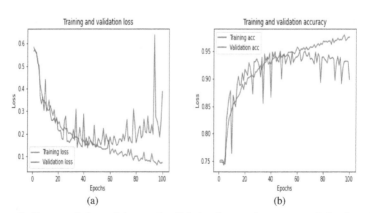

Figure 6: Curves of the training and validation loss and accuracy of the first dataset (fold 2).

3.3 Evaluation on Second Dataset

Similarly, this section presents in detail the results of applying the proposed model. In particular, WCE frames exhibiting red lesion regions and ones without any abnormality were utilized. These images are extracted from videos issued after a capsule endoscopy procedure. Table 4 shows the training and validation accuracies and losses obtained in the second dataset besides, the accuracy achieved in the testing dataset. The values are given for each fold. The proposed model achieved an average of 99.79% and 98.75% accuracy in the training and validation datasets, respectively. In terms of loss, the proposed approach achieved an average of 0.0063% and 0.0431% in the training and validation datasets, respectively. Besides, it reached 99.42% (0.0161) compared to 99.34% (0.0318), achieved by the Relu activation function, in the accuracy (loss) metric. In addition, Fig. 7 exhibits the corresponding training and validation curves for the loss and accuracy metrics obtained in the second dataset. These curves depict the history of training and validation of the third fold.

Table 4: Performance of the proposed method on second dataset.

Fold	Training accuracy (loss)	Validation accuracy (loss)	Testing accuracy (loss)	Testing accuracy (loss) (Relu)
Fold 1	0.9954 (0.0135)	0.9862 (0.0462)		
Fold 2	0.9990 (0.0031)	0.9896 (0.0371)	0.9942 (0.0161)	0.9934 (0.0318))
Fold 3	0.9993 (0.0025)	0.9869 (0.0461)		

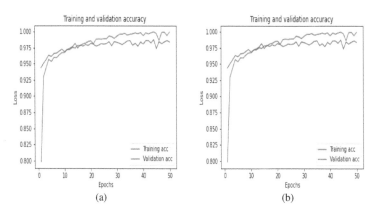

Figure 7: Curves of the training and validation loss and accuracy of the second dataset (fold 3).

4 Conclusion

In this chapter, we presented a computationally simple deep learning CNN model for ulcer and red lesion diseases classification. Besides, we have experienced the performance of PRenu activation function and compared it to Relu one. The Renu activation function outperformed the Relu one by 0.31% and 0.08%, in accuracy metric, in the first and second test datasets, respectively. In terms of loss, the PRenu reached 0.1085 (0.0161) compared to 0.1879 (0.0318) achieved by Relu in the first (second) test dataset, respectively. In addition, the presented model has been assessed on two different datasets and proved its efficiency. The proposed method has achieved 96.51%, 99.42% in terms of accuracy, in the first and second test datasets, respectively. In our future work, we will be addressing the problem of segmentation of ulcerous and red lesion regions. It will also converge towards generalizing the proposed framework for other medical imaging classification problems.

Acknowledgments

We gratefully acknowledge and express our thanks to the National Center for Scientific and technical Research (CNRST) in Rabat for its research and financial support under the Al-Khawarizmi program in the field of Artificial Intelligence and its Applications.

References

Alexis Eid, Vasileios S. Charisis, Leontios J. Hadjileontiadis and George D. Sergiadis. (2013). A curvelet-based lacunarity approach for ulcer detection from wireless capsule endoscopy images. In *Proceedings of the 26th IEEE International Symposium on Computer-Based Medical Systems*, pp. 273–278. IEEE.

Anjany Kumar Sekuboyina, Surya Teja Devarakonda and Chandra Sekhar Seelamantula. (2017). A convolutional neural network approach for abnormality detection in wireless capsule endoscopy. In *2017 IEEE 14th International Symposium on Biomedical Imaging (ISBI 2017)*, pp. 1057–1060. IEEE.

Baopu Li and Max Q. -H. Meng. (2009). Texture analysis for ulcer detection in capsule endoscopy images. *Image and Vision Computing*, 27(9): 1336–1342.

Edward J. Ciaccio, Suzanne K. Lewis, Govind Bhagat and Peter H. Green. (2017). Coeliac disease and the video capsule: What have we learned till now. *Annals of Translational Medicine*, 5(9).

Emanuele Rondonotti. (2017). Capsule retention: Prevention, diagnosis and management. *Annals of Translational Medicine*, 5(9).

Husanbir Singh Pannu, Sahil Ahuja, Nitin Dang, Sahil Soni and Avleen Kaur Malhi. (2020). Deep learning based image classification for intestinal hemorrhage. *Multimedia Tools and Applications*, 79(29): 21941–21966.

Ilyas El Jaafari, Ayoub Ellahyani and Said Charfi. (2021). Parametric rectified nonlinear unit (prenu) for convolution neural networks. *Signal, Image and Video Processing*, 15(2): 241–246.

Jia-sheng Yu, Jin Chen, Xiang, Z.Q. and Yue-Xian Zou. (2015). A hybrid convolutional neural networks with extreme learning machine for WCE image classification. In *2015 IEEE International Conference on Robotics and Biomimetics (ROBIO)*, pp. 1822–1827. IEEE.

Jinn-Yi Yeh, Tai-Hsi Wu, Wei-Jun Tsai et al. (2014). Bleeding and ulcer detection using wireless capsule endoscopy images. *Journal of Software Engineering and Applications*, 7(05): 422.

Jun-Yan He, Xiao Wu, Yu-Gang Jiang, Qiang Peng and Ramesh Jain. (2018). Hookworm detection in wireless capsule endoscopy images with deep learning. *IEEE Transactions on Image Processing*, 27(5): 2379–2392.

Kundu, A. K. and Fattah, S. A. (2017). An asymmetric indexed image based technique for automatic ulcer detection in wireless capsule endoscopy images. In *2017 IEEE Region 10 Humanitarian Technology Conference (R10-HTC)*, pp. 734–737. IEEE.

Kundu, A. K., Arnab Bhattacharjee, Fattah, S. A. and Shahnaz, C. (2016). Automatic ulcer detection scheme using gray scale histogram from wireless capsule endoscopy. In *2016 IEEE International WIE Conference on Electrical and Computer Engineering (WIECON-ECE)*, pp. 242–245. IEEE.

Kundu, A. K., Arnab Bhattacharjee, Fattah, S. A. and Shahnaz, C. (2017). An automatic ulcer detection scheme using histogram in YIQ domain from wireless capsule endoscopy images. In *TENCON 2017-2017 IEEE Region 10 Conference*, pp. 1300–1303. IEEE.

Lanmeng Xu, Shanhui Fan, Yihong Fan and Lihua Li. (2018). Automatic polyp recognition of small bowel in wireless capsule endoscopy images. In *Medical Imaging 2018: Imaging Informatics for Healthcare, Research, and Applications*, volume 10579, page 1057919. International Society for Optics and Photonics.

Lecheng Yu, Pong C. Yuen and Jianhuang Lai. (2012). Ulcer detection in wireless capsule endoscopy images. In *Proceedings of the 21st International Conference on Pattern Recognition (ICPR2012)*, pp. 45–48. IEEE.

Lequan Yu, Hao Chen, Qi Dou, Jing Qin and Pheng Ann Heng. (2016). Integrating online and offline three-dimensional deep learning for automated polyp detection in colonoscopy videos. *IEEE Journal of Biomedical and Health Informatics*, 21(1): 65–75.

Meryem Souaidi and Mohamed El Ansari. (2019). Multi-scale analysis of ulcer disease detection from WCE images. *IET Image Processing*, 13(12): 2233–2244.

Mohsen Hajabdollahi, Reza Esfandiarpoor, Elyas Sabeti, Nader Karimi, Reza Soroushmehr, S. M. et al. (2020). Multiple abnormality detection for automatic medical image diagnosis using bifurcated convolutional neural network. *Biomedical Signal Processing and Control*, 57: 101792.

Muhammad Sharif, Muhammad Attique Khan, Muhammad Rashid, Mussarat Yasmin, Farhat Afza et al. (2021). Deep cnn and geometric features-based gastrointestinal tract diseases detection and classification from wireless capsule endoscopy images. *Journal of Experimental & Theoretical Artificial Intelligence*, 33(4): 577–599, 2021.

Nimisha Elsa Koshy and Varun P. Gopi. (2015). A new method for ulcer detection in endoscopic images. In *2015 2nd International Conference on Electronics and Communication Systems (ICECS)*, pp. 1725–1729. IEEE.

Omid Haji Maghsoudi and Mahdi Alizadeh. (2018). Feature based framework to detect diseases, tumor, and bleeding in wireless capsule endoscopy. *arXiv preprint arXiv:1802.02232*.

Paulo Coelho, Ana Pereira, Marta Salgado and António Cunha. (2018). A deep learning approach for red lesions detection in video capsule endoscopies. In *International Conference Image Analysis and Recognition*, pp. 553–561. Springer.

Piotr Szczypiński, Artur Klepaczko, Marek Pazurek and Piotr Daniel. (2014). Texture and color based image segmentation and pathology detection in capsule endoscopy videos. *Computer Methods and Programs in Biomedicine*, 113(1): 396–411.

Prabhananthakumar Muruganantham and Senthil Murugan Balakrishnan. (2021). A survey on deep learning models for wireless capsule endoscopy image analysis. *International Journal of Cognitive Computing in Engineering*, 2: 83–92.

Ronald A. Castellino. (2005). Computer aided detection (cad): An overview. *Cancer Imaging*, 5(1): 17.

Rongsheng Zhu, Rong Zhang and Dixiu Xue. (2015). Lesion detection of endoscopy images based on convolutional neural network features. In *2015 8th International Congress on Image and Signal Processing (CISP)*, pp. 372–376. IEEE.

Ruwan Nawarathna, Jung Hwan Oh, Jayantha Muthukudage, Wallapak Tavanapong, Johnny Wong et al. (2014). Abnormal image detection in endoscopy videos using a filter bank and local binary patterns. *Neurocomputing*, 144: 70–91.

Şaban Öztürk and Umut Özkaya. (2021). Residual lstm layered cnn for classification of gastrointestinal tract diseases. *Journal of Biomedical Informatics*, 113: 103638.

Said Charfi and Mohamed El Ansari. (2017). Computer-aided diagnosis system for ulcer detection in wireless capsule endoscopy videos. In *2017 International Conference on Advanced Technologies for Signal and Image Processing (ATSIP)*, pp. 1–5. IEEE.

Said Charfi and Mohamed El Ansari. (2018). Computer-aided diagnosis system for colon abnormalities detection in wireless capsule endoscopy images. *Multimedia Tools and Applications*, 77(3): 4047–4064.

Said Charfi and Mohamed El Ansari. (2020). A locally based feature descriptor for abnormalities detection. *Soft Computing*, 24(6): 4469–4481.

Said Charfi, Mohamed El Ansari and Ilangko Balasingham. (2019). Computer-aided diagnosis system for ulcer detection in wireless capsule endoscopy images. *IET Image Processing*, 13(6): 1023–1030.

Santi Seguí, Michal Drozdzal, Guillem Pascual, Petia Radeva, Carolina Malagelada et al. (2016). Generic feature learning for wireless capsule endoscopy analysis. *Computers in Biology and Medicine*, 79: 163–172, 2016.

Shelly Soffer, Eyal Klang, Orit Shimon, Noy Nachmias, Rami Eliakim et al. (2020). Deep learning for wireless capsule endoscopy: A systematic review and meta-analysis. *Gastrointestinal Endoscopy*, 92(4): 831–839.

Shipra Suman, Fawnizu Azmadi Hussin, Walter Nicolas and Aamir Saeed Malik. (2016). Ulcer detection and classification of wireless capsule endoscopy images using rgb masking. *Advanced Science Letters*, 22(10): 2764–2768.

Shuai Wang, Yang Cong, Hancan Zhu, Xianyi Chen, Liangqiong Qu et al. (2020). Multi-scale context-guided deep network for automated lesion segmentation with endoscopy images of gastrointestinal tract. *IEEE Journal of Biomedical and Health Informatics*, 25(2): 514–525.

Sindhu, C. P. and Vysak Valsan. (2017). Automatic detection of colonic polyps and tumor in wireless capsule endoscopy images using hybrid patch extraction and supervised classification. In *2017 International Conference on Innovations in Information, Embedded and Communication Systems (ICIIECS)*, pp. 1–5. IEEE.

Tariq Rahim, Muhammad Arslan Usman and Soo Young Shin. (2020). A survey on contemporary computer-aided tumor, polyp, and ulcer detection methods in wireless capsule endoscopy imaging. *Computerized Medical Imaging and Graphics*, pp. 101767.

Tien-Chun Chang. (2015). The role of computer-aided detection and diagnosis system in the differential diagnosis of thyroid lesions in ultrasonography. *Journal of Medical Ultrasound*, 23(4): 177–184.

Tonmoy Ghosh, Antara Das and Rosni Sayed. (2016). Automatic small intestinal ulcer detection in capsule endoscopy images. *International Journal of Scientific and Engineering Research*, 7(10): 737–741.

US Food, Drug Administration et al. (2012). Computer-assisted detection devices applied to radiology images and radiology device data—premarket notification [510 (k)] submissions. *Silver Spring: Food and Drug Administration*.

Vasileios S. Charisis, Christina Katsimerou, Leontios J. Hadjileontiadis, Christos N. Liatsos and George D. Sergiadis. (2013). Computer-aided capsule endoscopy images evaluation based on color rotation and texture features: An educational tool to physicians. In *Proceedings of the 26th IEEE International Symposium On Computer-Based Medical Systems*, pp. 203–208. IEEE.

Vasileios S. Charisis, Leontios J. Hadjileontiadis, Christos N. Liatsos, Christos C. Mavrogiannis and George D. Sergiadis. (2012). Capsule endoscopy image analysis using texture information from various colour models. *Computer Methods and Programs in Biomedicine*, 107(1): 61–74.

Vasileios S. Charisis, Leontios J. Hadjileontiadis, João Barroso and George D. Sergiadis. (2012). Intrinsic higher-order correlation and lacunarity analysis for wce-based ulcer classification. In *2012 25th IEEE International Symposium on Computer-Based Medical Systems (CBMS)*, pp. 1–6. IEEE.

Xiao Jia and Max Q. -H. Meng. (2016). A deep convolutional neural network for bleeding detection in wireless capsule endoscopy images. In *2016 38th Annual International Conference of the IEEE Engineering in Medicine and Biology Society (EMBC)*, pp. 639–642. IEEE.

Xiaoqing Guo and Yixuan Yuan. (2020). Semi-supervised wce image classification with adaptive aggregated attention. *Medical Image Analysis*, 64: 101733.

Yixuan Yuan and Max Q. -H. Meng. (2017). Deep learning for polyp recognition in wireless capsule endoscopy images. *Medical Physics*, 44(4): 1379–1389.

Yixuan Yuan, Baopu Li and Max Q. -H. Meng. (2016). Wce abnormality detection based on saliency and adaptive locality-constrained linear coding. *IEEE Transactions on Automation Science and Engineering*, 14(1): 149–159.

Yixuan Yuan, Jiaole Wang, Baopu Li and Max Q. -H. Meng. (2015). Saliency based ulcer detection for wireless capsule endoscopy diagnosis. *IEEE Transactions on Medical Imaging*, 34(10): 2046–2057.

Yuexian Zou, Lei Li, Yi Wang, Jiasheng Yu, Yi Li et al. (2015). Classifying digestive organs in wireless capsule endoscopy images based on deep convolutional neural network. In *2015 IEEE International Conference on Digital Signal Processing (DSP)*, pp. 1274–1278. IEEE.

Zahra Amiri, Hamid Hassanpour and Azeddine Beghdadi. (2022). Feature extraction for abnormality detection in capsule endoscopy images. *Biomedical Signal Processing and Control*, 71: 103219.

Chapter 6

Do More With Less
Deep Learning in Medical Imaging

Shivani Rohilla,* Mahipal Jadeja and *Emmanuel S. Pilli*

ABSTRACT

The emergence of deep learning (DL) has the power to transform the entire picture of healthcare, and it has been used diligently to detect diseases. Medical imaging is one of those areas where a set of techniques create visual representations of the internal parts of the body. By using deep learning, one can do detection and prediction effectively and efficiently. In the healthcare domain, there is a need to locate anomalies and recognize specific indications of different diseases. This chapter focuses on analytic methods that have used deep-learning techniques in medical imaging. The chapter also covers the most common challenges incurred in deep learning based medical imaging solutions and suggests possible solutions. It emphasizes how deep learning with image classification and segmentation is applied in medical imaging. Further, a case study of cardiovascular disease detection using a convolution neural network along with some existing works in the literature is presented. The chapter serves as a basic building block for deep learning in medical applications, and it offers viewpoints to further investigate the domain over more real-world problems. It also highlights the future scope and research directions in this area.

Malaviya National Institute of Technology, Jaipur, Rajasthan.
Emails: mahipaljadeja.cse@mnit.ac.in; espilli.cse@mnit.ac.in
* Corresponding author: 2019rcp9560@mnit.ac.in

1 Introduction

This introduction discusses what deep learning, neural networks, and medical imaging are. Various deep learning techniques used for imaging will also be discussed. After examining these basic concepts, the chapter highlights the deep learning algorithms and how DL is a critical player in healthcare.

1.1 Relevance of Deep Learning

In the 1940s, people struggled with problematic computational tasks; the term deep learning was nonetheless coined (https://towardsdatascience.com/the-deep-history-of-deep-learning-3bebeb810fb2 last accessed on 12/09/2021). No one was aware of the deep learning concept a few years back because the phase of applying human intelligence was at its peak. As soon as Artificial Intelligence came into the picture, machine learning (ML) as well as deep learning (DL) were introduced. DL outdoes ML-based techniques and algorithms in performance and is mainly used to accomplish tasks. Deep learning is used widely nowadays because it improves the system's performance while learning or predicting tasks. Although DL has grown a lot after the 90s, its necessary to know where and how it has grown over time.

The deep learning market is predicted to perform an annual growth rate of 42.56% from 2020 to 2025 (https://www.mordorintelligence.com/industry-reports/deep-learning last accessed on 12/09/2021). Factors leading to the growth of an enterprise and efficient performance have convinced companies to invest in DL-based solutions. At the same time, ML approaches with their new hybrid techniques started working well. These approaches offer better accuracies. Some novel classes of neural networks are also developed. Figure 1

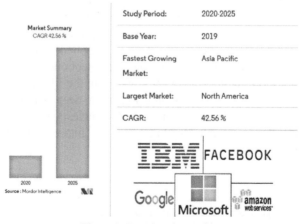

Figure 1: Deep learning market.

shows the market snapshot of deep learning (https://www.mordorintelligence.com/industry-reports/deep-learning last accessed on 12/09/2021).

Nowadays, the healthcare sector is transforming from paper-based records to electronic health records (EHR). This has opened the gate of opportunities for the application of deep learning (https://www.analyticsvidhya.com/blog/2021/08/deep-learning-in-health-care-a-ray-of-hope-in-the-complex-medical-world last accessed on 12/09/2021). It will improve patient-clinician interactions, thereby resulting in improved health outcomes in a hospital setup. After understanding these preliminaries, we can define Deep Learning as (https://www.ibm.com/cloud/learn/deep-learning last accessed on 12/09/2021): *"Deep learning is a subset of machine learning, which is essentially a neural network with three or more layers. These neural networks attempt to simulate the behaviour of the human brain—albeit far from matching its ability—allowing it to "learn" from large amounts of data."*

A basic neural network with only three layers can fulfil the task, but still, effectiveness is absent. A number of layers are required to perform the task dexterously. Deep learning makes use of layered architecture. Data is processed through a number of layers, mapping is done, and the next layer retrieves the information from the previous layer's output. The confirmation is done in the same way. In this way, connections are made between the multiple layers, and the feature of deep learning to purify the correlation process is strengthened. To understand deep learning, we must first have a rough idea of how neurons travel in the uppermost part of our body, i.e., the brain. It is loosely related to this aspect. Electrical impulses travel across the brain; nodes in each layer are activated after receiving stimulus from their neighbour neurons. The same is with the hierarchical layers of the DL network.

AI applications that make use of automation are improving day by day with the help of DL. DL-based technology may be applied to digital assistants, electronic devices using VoIP, fully automatic vehicles, and many other daily use products. Gradually, deep learning is going to be a part of our daily lives. We can now discuss some exciting insights and analyses on the deep neural network techniques and algorithms and their application in medical imaging by understanding the whole idea.

1.2 Building Blocks of a Neural Network

Neural networks or artificial neural networks (ANNs) are feed-forward networks that assign a task to a specific portion of a layer in the network, and traversing is done from left to right. After so many passes, the output is refined. Many a time, we get the optimal result in a very short time. We can

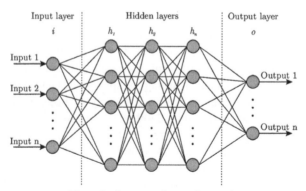

Figure 2: Structure of a neural network.

call it the best case of ANN. These layers are hidden in the network and are assigned to perform the mathematical work (Fig. 2).

Neural networks (NN) were made to design the patterns, work with them, and produce the output according to human needs. NN elucidate data from actuators using machine knowledge, prediction, labeling, or extracting raw data. Here comes the idea of pattern recognition. The patterns they recognize are in the form of mathematical vectors or images, text, sound, among others. It is then converted into a specific form to understand and get the desired output. Deep learning is sometimes called stacked, hierarchical, or layered neural network but with a pinch of more intelligence than other learning techniques for specific problems (https://wiki.pathmind.com/neural-network last accessed on 12/09/2021).

The layers are made of nodes where computation occurs. The function of a node is to assign a weight and start intensifying the input. Importance has been given to each node according to its specified task. For example, the task of classification/segmentation with minimal loss. The concatenation of these input weights is summed up and mapped using an activation function. At this stage, it is decided to amplify or dampen the signal. If we allow the signal to amplify, the neuron comes into action (Fig. 3).

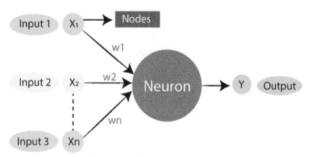

Figure 3: Structure of a node.

1.3 Deep Learning—"A Boon for Healthcare"

While the application of deep learning in some fields might seem ordinary right now, one field that cannot survive without it is healthcare. With the expansion of computing power after CPUs, i.e., with GPUs and the supply of large datasets, deep learning overtops other technologies. Artificial Intelligence, Machine learning, and Deep learning have supported medical imaging with minimum disturbance and confusion. The growth of Artificial Neural Networks (ANN) and Convolution Neural Networks (CNNs) in medical imaging can empower the whole ecosystem (Currie et al., 2019). After getting an insight into these three technologies, one can quickly build a fast and efficient model for medical imaging purposes. A need for successful and unbeatable implementation in medical imaging with AI techniques has led to the growth of DL and convolution neural networks (CNNs).

Overall, we need a technology that can outperform all other technologies in terms of accuracy, specificity, low loss, low error rate, and is easy to incorporate. Data has to be classified as well as segmented. It must reduce delays and increase efficiency. Also, it should be fast and cost-effective.

Currie (Currie, 2019) states that object detection, segmentation, and object classification are widespread deep learning applications. These operations are often outpoured and supplied to other medical applications. In automated radiation treatment planning, object segmentation can represent tumours as targets and organs as no dose regions. Object Detection is widely used in CADs of mammograms to diagnose tumours the same as in CT scans of other organs. It is true that deep learning smartly works with datasets for the diagnosis of a disease, object recognition, value prediction, and other automated tasks like feature extraction, classification. It will surely help medical practitioners to perform healthcare tasks in no time.

1.4 Convolution Neural Networks (CNNs) and Medical Imaging

The outputs which we get in the form of generic descriptors from CNNs are pretty helpful. They are used in object recognition as well as localization in the case of natural images. Medical analysis players across the globe are planning and performing tasks using DL-based techniques. Many applications of DL are coming into the picture. Out of them, medical imaging is considered in this chapter. For the diagnostic purpose, the following medical images are generally considered: X-ray images, Computed tomography scan (CT), Ultrasound Imaging, and MRI (Magnetic resonance imaging). We consider four tasks for medical image analysis which we have discussed so far, i.e., image classification, object detection, image segmentation, and registration. Image segmentation is the fundamental task in computer vision. Convolution neural networks dominate the process of classification among other networks. As deep learning is developing, the frameworks of CNN have also improved. AlexNet, VGGNet, denseNet are some of its examples. Following evaluation metrics are considered for the evaluation of image classification algorithms:

$$\text{Precision} = \frac{TP}{TP + FP},$$

$$\text{Recall} = \frac{TP}{TP + FN},$$

$$\text{Accuracy} = \frac{TP + TN}{n},$$

$$F_1 = 2 \cdot \frac{\text{Precision} \cdot \text{Recall}}{\text{Precision} + \text{Recall}}.$$

Object detection algorithms include both identification as well as localization tasks. In medical image analysis, the objective is to detect the signs of abnormality as early as possible. Some of the famous applications include detection of a lung nodule in the chest X-ray/CT scan images, lesion detection in the computed tomography images. The third one, i.e., image segmentation, determines anatomical structures present in images. Evaluation of segmentation methods is mainly based on the dice similarity coefficient and intersection over union (IOU).

$$\text{Dice} = \frac{2 \times TP}{2 \times TP + FP + FN},$$

$$\text{IOU} = \frac{TP}{TP + FP + FN}.$$

Medical Image registration maintains correspondence within images and is widely used in applications like a fusion of MRI or CT images with emission tomography techniques for disease detection and prevention.

Promising results are emerging because the accuracy of diagnosis is not compromised yet. Image acquisition has greatly improved in the past decade, with devices acquiring data faster and an increased aspect ratio. The image interpretation process, however, has begun to benefit from emerging technologies. Although physicians, medical practitioners, and doctors interpret the results well, humans sometimes fail due to their subjectivity, differences in the interpreters, and long working hours. Various diagnosis cases need a preliminary search test and detection of abnormalities. Computerized tools are responsible for improving diagnosis by enabling identification of the tests that require treatment. This supports the workflow of deep learning enthusiasts. Deep learning draws a new boundary of data analytics with magical results never experienced in the past decade.

This chapter is divided into five major sections. Starting from the first section, i.e., the introduction of deep neural networks, it moves towards various deep learning techniques in medical imaging. The third section shows a case of disease detection and prediction and original results drawn by a few researchers in the past three years. The fourth section points out the challenges incurred and their possible solutions. Last but not least, the fifth section of this chapter throws light on 2021's open research issues and directions in the field of DL in healthcare.

2 Deep Learning Techniques for Segmentation of Medical Images

Spotting the pixels corresponding to a biological part in the medical images, e.g., MRI images, is known as Medical Image Segmentation (MIS). CT scans, MRIs and EEGs have proven to be critical in terms of drawing information in the diagnosis process. A lot of research work has been done in medical imaging in the past few years. Researchers are making use of different technologies to boost up the process of diagnosis and prediction. Mathematical methods and filters were used in the old times to get results. However, as the technology goes hand in hand, we are improving in terms of performance. Artificial intelligence gave birth to machine learning, and machine learning grabbed the market. Designing, extraction, and prediction were all made easy with machine learning. People started using its applications. Researchers started inventing new hybrid techniques and novel algorithms. Accuracy was one of the main problems. In the early 2000s, deep learning approaches started to exemplify their new potentials in imaging tasks. The uniqueness and less complex structure of deep learning has made it a popular technique for medical imaging. Shen et al. (Shen et al., 2017) demonstrated various types of medical image analysis without focusing on the practical aspect of image segmentation. Litjens et al., discussed registration, classification, and detection tasks.

In this section, the basics of major techniques related to deep learning-based medical imaging are discussed. Along with the explanation of these techniques, relevant state-of-the-art research has been mentioned.

2.1 Convolutional Neural Networks (CNNs)

A CNN is an algorithm in deep learning that uses several hidden layers followed by mapping, pooling, and many other tasks. The basic objective of shifting from ANN to CNN is because of its unique capability to handle images. CNNs work best with images. Be it MRI, EEG, ECG, CT scan, or any other medical imaging data, CNN is always one of the best options. Each branch of the layer in CNN performs its designated task like loss calculations and augmentation. One layer produces the output, which acts as an input for the next layer (Fig. 4). The first layer is always connected to the input image. It is considered the beginning layer of the hierarchy. The number of pixels is considered as neurons here. Another set comprises convolutional layers responsible for convolution and mapping different filters with the input data. This layer itself extracts the features. Filters are applied by the designers based upon their size. They are also called kernels. Every neuron gives an answer to a specific area known as the receptive field, and the output is known as activation mapping. It signifies the effect of a particular filter passed into the network. The activation layers are usually applied after convolution layers so that non-linear operations can be performed on the activation maps. Pooling is another method to improve the performance of this system. It can be done to reduce the dimensionality of the output drawn after convolution. Max or average pooling can be applied according to the severity of the problem. At last, fully connected layers are used for obtaining high-level abstractions. Back propagation is used in the training phase in order to obtain optimized weights for the underlying neural network as well as kernels (Commandeur et al., 2018). Now, we consider the popular types of CNNs.

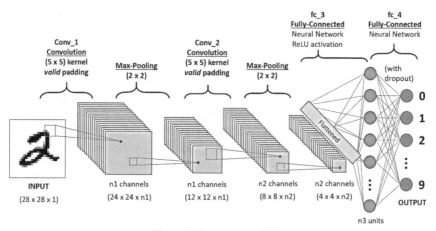

Figure 4: Structure of a CNN.

2.1.1 2D CNN

The performance of a CNN in pattern recognition has proven to be effective. Therefore, seeking CNN for medical image segmentation has been demonstrated by more researchers. In 2D convolution, we proceed with the segmentation task using the 2D image as input and passing it with 2D filters. Zhang et al. (Zhang et al., 2015) stated that information is fed to the input layer of the network. R, G, B image channels are mentioned here. These sources of information are 2D images. It helps to find out if multimodality helps to improve the results. The results contain better metrics than those with a single modality input. Bar et al. (Bar et al., 2015) proposed a transfer learning-based approach in which a pre-trained model gives low-level features to ImageNet. PiCoDes (Bergamo et al., 2011) provides high-level features, and then all features are merged. This 2D convolution neural network gives better results in medical imaging. 2.5 D is another way of doing it with some variations.

2.1.2 3D CNN

2D kernels are the only way to process in 2D or 2.5 D convolution cases. They cannot be applied to 3D filters. The significance of the 3D filter is to draw a robust visualization in terms of 3D representation with the help of axes- x, y, and z. The structure of 3D CNN is very much similar to 2D except for the need to apply 3D modules at critical points in the network, i.e., with 3D layers for convolution and sub-sampling layers. The structure is trained to point out the central voxels and works on them.

The availability of 3D medical imaging recommends the use of 3D information for segmentation. Kleesiek et al. (Kleesiek et al., 2014) stated that they had used 3D models for the segmentation process of brain MRI images. Their idea was applied by Kamnitsas et al. (Kamnitsas et al., 2015) to develop a dual-path 3D CNN in which 2 parallel pathways were considered. Sub-sampled image representation was considered by the next pathway. This helps in processing large portions surrounding the voxel, which allows the complete system to multi-scale. This variation while using a 3 x 3 filter has given better accuracy, sensitivity, and less processing time than its original design. Dou et al. [23] mentioned a few 3D kernels which spatially divide the weights and help to decrease the total number of parameters. While working with images and applying 3D filters, 3D modelling of images is challenging. For the sub-sampling layer, a few other techniques can be considered, like 3D max pooling. But once done, it helps to attain a faster-converging speed as compared to pure 3D CNN. A simple structure is shown in Fig. 5.

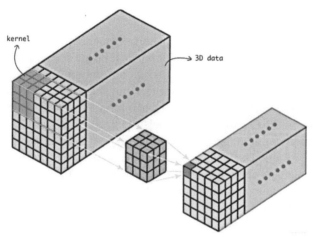

Figure 5: 3D convolution neural network.

2.2 Fully Convolution Networks (FCN)

Long et al. (Long et al., 2015) demonstrated the fully convolution layer (Fig. 6). A dense pixel-wise prediction can be made using this technique. To bring off a better localization performance, they took the help of activation maps with a nice resolution, joined them with up-sampled outputs, and passed them to the convolution layers. This helped in achieving better accuracy. This process's improvements allow the FCN to predict pixel by pixel images from the complete image rather than patch by patch prediction. It performs the prediction in a single pass; therefore, it is known as FCN. Nie et al. (Nie et al., 2016) have also done the same experiment. The results indicate the dominance of FCN over convolution neural networks since it offers a 0.885 mean dice coefficient value. It was previously 0.864. Fully convolution networks can further be cascaded, faded, and multi-streamed.

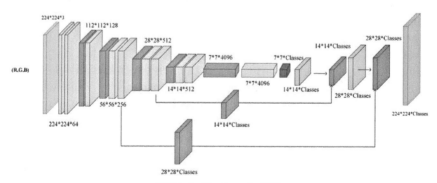

Figure 6: Structure of a FCN.

2.3 U-Net

In healthcare, apart from diagnosing a disease, detecting the area of abnormality is also to be considered. For this, U-Net was discovered. Ronneberger et al. (Ronneberger et al., 2015) proposed U- Net. Zeiler and Fergus (Zeiler and Fergus, 2014) discussed the term deconvolution.

Taking an idea from the building blocks of FCN, U-Net was developed. U-Net has a lovely pattern of skip connections between various layers and stages of the network (Christ et al., 2016). It has twenty-three layers that form its structure. Some big patches result in less accuracy, so there are two paths. One is the analysis, and the other is synthesis. The analysis path obeys the architecture of the convolution neural network (see Fig. 7). The structure of the synthesis path contains an up-sampling layer before the deconvolution layer. The unique quality of U-Net is doing image localization using pixel by pixel image prediction. It can further be categorized as 3D U-net and V-net, which are slightly different networks.

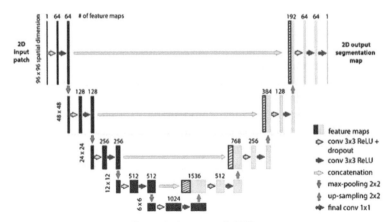

Figure 7: Structure of a U-Net.

2.4 Convolutional Residual Networks (CRNs)

Deep neural networks are much better for learning, but as the depth increases, the accuracy is affected. He et al. (He et al., 2016) stated residual networks are a novelty for natural image segmentation for 2D imaging. They took an idea from a deeper network structure. Here, a residual map is passed to some layers only. The stacked layers are not fed. The design of this network is better to gain maximum accuracy (Fig. 8). Yu et al. (Yu et al., 2016) contributed to the design and implementation of CRN for melanoma recognition. The key benefit of FCRN over simple CRN is: it can do predictions pixel by pixel, which may be an essential attribute for many segmentation tasks.

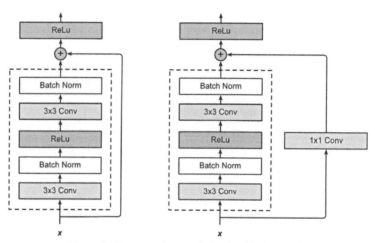

Figure 8: Structure of a convolutional residual network.

Kawahara et al. (Kawahara et al., 2016) consider both local and global contextual features to gain advantages over other deep neural models. Here, the model is enhanced by multi-scaling, resulting in a completely deeper FCRN consisting of fifty layers. 2D deep residual networks are used in many medical image segmentation tasks (Gibson et al., 2017) and image processing tasks (He et al., 2016; Zagoruyko and Komodakis, 2016), yet few studies have applied CRN algorithms. VoxResNet, by Chen et al. (Chen et al., 2018), is a 3D DRN with a design of a 25-layer model fed to 3D MRI brain images. Tiny convolutional kernels are applied in (Simonyan and Zisserman, 2014). A big receptive field and more information were taken as outputs as 3 convolutional layers with a march of two were built. This reduces the resolution by a factor of 8. Also, VoxResNet has been extended to auto-context VoxResNet. Auto-context learns an integrated low-level, context-based model. This generic model is easy to implement. It can process images with two or more modalities for gaining more powerful segmentation. The work done using this network shows that there is always a good result of using more deep hidden layers. Although identification of data is always necessary, but to improve the performance, we need to dig deeper.

2.5 Recurrent Neural Networks

The recurrent neural networks (RNNs) have recurrent connections. These special connections help the network to remember the last input patterns. There are many slices, one after the other, in the region of interest pooling (ROI) which outputs the correlations in the following slices. RNNs can extract sequential data from the slices given in the input set.

Various types of RNNs are long short-term memory (LSTM), Gated Recurrent Unit (GRU), and clockwork RNN (CW-RNN). Among these, LSTM is explained by Hochreiter and Schmidhuber (Hochreiter and Schmidhuber, 1997) and is identified as a unique type of RNN. If we talk about the LSTM network, inputs are vectorized, which is not an advantage in terms of image segmentation tasks as spatial information cannot be drawn. So, convolutional long short-term memory (CLSTM) (Srivastava et al., 2015; Xingjian et al., 2015) is a good option where only convolution is done. It does not vectorize the units. LSTM can be contextual too.

3 Case Study: Arrhythmia detection using Deep Learning Techniques

Cardiovascular infections are one of the major causes of death across the globe. CVD accounts for 17.3 million deaths each year around the world. It is estimated that it can be greater than 23.6 million by 2030. Despite progress, 20–40% of heart strokes and attacks capture the people who have been previously undiagnosed with any heart problem. The quality of life of patients after recovering from heart disease is also not taken into account. Only 5% of patients with severe heart failure receive palliative care. There is a high time to address this issue. We must know that a step of research or discovery may save a life in no time.

Let's take a case study of Arrhythmia. Cardiac Arrhythmia is a serious heart condition that occurs when electric signals are disturbed and start fluctuating (Fig. 9). It is very unsure that a patient may have symptoms, or sometimes internally, he/she is suffering from cardiac Arrhythmia. The patient needs a quick diagnosis in no time. Deep learning algorithms can do a lot to reverse worrisome recent trends by implementing deep learning models to predict the disease quickly and efficiently.

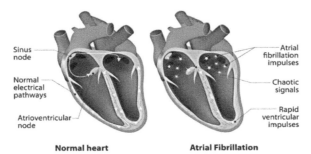

Figure 9: Arrhythmia—A CVD.

Here, we are showing a comparison of various deep learning models practically implemented for detecting Arrhythmia: a condition leading to cardiovascular disease (from ECG recordings) is given below:

Table 1: Comparison metrics of various research works in the last three years.

S. no.	Author	Year	Dataset	Technique	Accuracy	Sensitivity	Specificity	F1 score
1	Erdenebayar et al.	2019	SA (sleep apnea)	CNN	98.7	98.6	98.7	–
2	Ulas Baran et al.	2019	PTB	CNN	99.78	–	–	–
3	Hamido Fujita et al.	2019	MIT-BIH	CNN	98.45	99.27	99.87	–
4	Rasmus et al.	2019		CNN+RNN		98.98	96.95	–
5	Hu et al.	2019	MIT-BIH	CNN	94.44	93.33	98.33	–
6	Isler et al.	2019	MIT-BIH	CNN	98.80	100	98.1	–
7	Wang et al.	2019	MIT-BIH	CNN	87.54	76.71	99.22	–
8	Ahmet et al.	2019	MIT-BIH	Hybrid Alexnet+SVM	96.77 & 90.67 with LSTM	–	–	–
9	Tanvir Mahmud et al.	2019	MIT-BIH	CNN	99.28	99.13	99.08	98.11
10	Jingshan Huang et al.	2019	MIT-BIH	1D+2D CNN	90.93, 99 resp.	–	–	–
11	Elif et al.	2019	MIT-BIH	2D CNN	97.42	–	–	–
12	Saroj et al.	2019	MIT-BIH	CNN+SMOTE	98.3	–	–	–
13	Wenjuan Cai et al.	2020	China physiological signal challenge 2018 data	DDNN	99.35	99.19	99.44	–
14	Rohan et al.	2020	MIT	CNN	99.03	99.800	81.25	–
15	Xue Xiang et al.	2020	MIT	CNN	99.43	97.5	–	–
16	Shuren Zhau et al.	2020	MIT	CNN+ELM	97.5	–	–	–
17	Chen Chen et al.	2020	MIT	CNN+LSTM	99.32	–	–	–
18	Georgios et al.	2021	MIT	CNN+LSTM	–	97.87	99.29	–
19	Ahmed et al.	2021	MIT	CNN+CQ-NSGT	99.84	–	–	–

4 Challenges with Training Deep Models

Deep learning techniques can be of great use when we have a large dataset. Sometimes unavailability occurs, and we can face the issues like effectiveness, relevance, and reliance on learned features or combining them with locally crafted features for modelling. Besides these, there are some other challenges too:

- **Overfitting**

Overfitting occurs when we have the highest accuracy of 98–100%, or we have an original dataset with which our training data matches completely. The accuracy of the model is relatively higher than other instances which have not been processed yet (Hwang and Kim, 2016). If we have a very small dataset for training, overfitting may occur soon. So, we need to apply such techniques or implement ways that can increase the amount of our data (Shen et al., 2017). Data augmentation is a nice technique to solve the problem of overfitting. We load more data for training here (Feng et al., 2017). Dropout can also be applied to overcome overfitting by discarding the layers which signify the background of the image (Srivastava et al., 2014). Drop connect is another way out to overcome dropout (Roth et al., 2014).

- **Training time**

To achieve fast convergence, one has to decrease the training time, which may be a difficult task at times. Here, we can apply pooling layers. It reduces the size of dimensions. When the size decreases, training time automatically becomes lesser (Dou et al., 2017). Some pooling-based solutions use convolution with stride (Simonyan and Zisserman, 2014). They have a good effect and solve the problem by lightening the network. Batch normalization standardizes inputs of a network to accelerate training. It reduces the generalization error and provides a bit of regularization, too (Ioffe and Szegedy, 2015). This is also called an effective key for fast convergence (Kawahara et al., 2016; Baumgartner et al., 2017; Çiçek et al., 2016). It is one of the best ways to boost network convergence as it does not harm the performance. However, in some cases, pooling can result in the loss of required information. Down sampling also falls in the same category.

- **Gradient vanishing**

Deeper networks always perform better but suffer from a gradient vanishing problem (Srivastava et al., 2014). The loss incurred cannot be paid back to the original input layers, which were shallow at times, mainly in 3D convolution.

If one wants to overcome the problem of gradient vanishing, a supervised approach is to be followed. Here, the output of intermediate layers is deconvoluted and passed to the SoftMax activation function. The losses of the hidden layer are stacked together to strengthen the gradient (Dou et al., 2017; Gibson et al., 2017; Zeng and Zheng, 2018).

- **Organ appearance**

The target body part may vary in terms of size, shape, and location, and it also differs from person to person (Codella et al., 2018). The human body is made up of diversified body organs. This is the biggest challenge during the application of any intelligent technique on the body. Deep learning is a complete package providing solutions for all types of organs. Increasing the deep layers of the network is has proved effective (Yu et al., 2016).

- **3D challenges**

If we consider the above challenges, it is much tougher to train volumetric data because of the low-voice variance between the targeted pixel or area and its neighbouring voxels. More parameters and a limited amount of training data remain. If we apply dense layers, it can suddenly decrease the inference period (Kleesiek et al., 2014).

- **Variability and reproducibility**

Deep learning is classical in terms of accuracy. However, its complexity results in variability. This variability is the reason for the reproducibility of results. The hyperparameters, datasets (observed versus real), the architecture, and a few other reasons may result in the challenge of variability in the model. Reproducibility in the model is often not recognized just because of this issue. In DL, due to the variability issues, one ignores the reproducibility. To overcome this problem, testing deep learning algorithms on different private/public datasets and the availability of data points should be taken into account. Other methods like cross-validation, and data augmentation may also be applied.

5 Open Research Issues and Directions

Data availability, effective deep learning techniques, and methods with processing power are the requisites that welcome deep learning uprisings. While the remarkable help of deep learning is completely noteworthy, so is the endeavor and cost of large models. Google DeepMind, IBS Watson, along with leading research labs, hospitals, and vendors, are collaborating for the development of an effective solution to medical imaging with large amounts of data. Siemens, Philips, Hitachi, and GE Healthcare have already made big investments.

Some directions and issues with their solutions are mentioned below:

- **Extensive inter-organization collaboration**

Stakeholders are waiting to replace humans with machines. Deep learning is doing great, so people should work by collaborating. This collaboration may be between machine learning scientists or service providers. This step can be taken to enhance the quality of healthcare. The togetherness will overcome the problem of lesser data availability. Also, the main issue can be the requirement of polished techniques to deal with huge healthcare data.

- **Work for image datasets**

Deep learning needs big datasets for efficient processing. Annotated data is not always available for work in this area, so we can co-operate with health care service providers by generating and sharing more real-time data. Grouping of image data from various sources may be hectic or expensive; therefore, this area may be good to work on. We can work for the availability of extremely large datasets for DL in our future research.

- **Advancement in deep learning methods**

We don't have big data available every time, so the supervised deep learning area is required to make a move from supervised to unsupervised/semi-supervised systems. We may not have medical data which is supervised at all times. Keeping in mind the accuracy of the DL model, we have to think about transform learning instead of supervised learning. Deep learning does not have exact solutions to this problem despite many efforts, and many questions are still unanswered because the opportunity for advancement is always knocking on the door.

- **Deep learning black box**

Interoperability is always an issue while we work for a big healthcare system. People don't always know each and every parameter, so they are insufficient in terms of information sometimes. Deep learning should be applied along with some other techniques too. Peeping inside the black box to find out what is inside is a major research issue, and deep learning enthusiasts are finding solutions for this problem. This may be a good direction for further research. Refer to Fig. 10 to understand a deep learning black box.

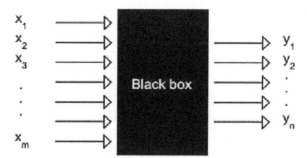

Figure 10: Deep learning: A black box.

- **Data privacy**

Data privacy is affected by sociological and technical issues. In order to address these issues, both sociological and technical perspectives need to be considered. The Health Insurance Portability and Accountability Act gives legal permission to healthcare system providers and stakeholders and applies a check on publicizing medical data. As the healthcare data is being increased, the privacy of patients' data is at stake. But HIPPA provides legal rights to patients to keep their data private. Furthermore, as real data is constantly increasing, there is a high need to increase efficient techniques and methods for its storage and processing.

- **Use of simulation environment**

The use of a simulation environment can be a good direction to work. We should use a simulation environment to work on real-world problems with real-time data. The benefit is we will be having good sources of information available in the domain.

- **Combining deep learning with knowledge systems**

Deep learning is itself efficient. When we talk about complex systems, they always need a fully automatic expert system. For increasing the quality of any deep learning model, we can integrate it with some expert system in the future.

- **Designing new algorithms**

The need for the best performance metrics is always a dream of every researcher. One can experiment with different techniques with each other in a single method. Designing hybrid deep learning models in the future may be a good option.

6 Conclusions and Future Scope

The chapter highlighted the fact that the benefits of deep learning in healthcare are abundant. These methods are fast, efficient, and accurate. Throughout the chapter, the most common challenges incurred are discussed, along with possible solutions. The articles emphasize how deep learning is being applied in medical imaging for classifying and segmenting images. This chapter tabulates the work of different researchers in the field of Arrhythmia detection, which compares the accuracy, sensitivity, scalability, and specificity of different CNN-based deep learning models. Deep learning-based medical imaging solutions can be used for different clinical purposes. Although deep learning models perform best in medical image analysis, still, there is a lack of benchmark datasets. This chapter has discussed the future scope of deep learning in the healthcare sector along with various research issues. The benefits that DL brings in the medical domain have been recognized by this chapter. The chapter can serve as a basic starting point for readers who want to work in deep learning-based medical imaging.

A new breed of image analysis software recently emerged using deep learning to eliminate a lot of repetitive, time-consuming tasks that medical practitioners perform regularly. With the growing array of products available for imaging-based diagnosis workflows, various steps can be automated. There is a lot of scope in terms of performance improvements for these deep learning-based products. Along with the CNN-based deep learning solutions, transfer learning may be applied to obtain better results. Federated learning can also be considered in which training is done in multiple decentralized devices.

References

AI Wiki, Homepage https://wiki.pathmind.com/neural-network last accessed on 12/09/2021.

Andersen, R. S., Peimankar, A. and Puthusserypady, S. (2019). A deep learning approach for real-time detection of atrial fibrillation. *Expert Systems with Applications*, 115: 465–473.

Baloglu, U. B., Talo, M., Yildirim, O., San Tan, R., Acharya, U. R. et al. (2019). Classification of myocardial infarction with multi-lead ECG signals and deep CNN. *Pattern Recognition Letters*, 122: 23–30.

Bar, Y., Diamant, I., Wolf, L. and Greenspan, H. (2015). Deep learning with non-medical training used for chest pathology identification. In *Medical Imaging 2015: Computer-Aided Diagnosis*, vol. 9414. International Society for Optics and Photonics, p. 94140V.

Baumgartner, C. F., Koch, L. M., Pollefeys, M. and Konukoglu, E. (2017). An exploration of 2d and 3d deep learning techniques for cardiac MR image segmentation. In *International Workshop on Statistical Atlases and Computational Models of the Heart*. Springer, pp. 111–119.

Bergamo, A., Torresani, L. and Fitzgibbon, A. W. (2011). Picodes: Learning a compact code for novel category recognition. In *NIPS*, 1(5). Citeseer, p. 6.

Cai, W., Chen, Y., Guo, J., Han, B., Shi, Y. et al. (2020). Accurate detection of atrial fibrillation from 12-lead ECG using deep neural network. *Computers in Biology and Medicine*, 116: 103378.

Chen, C., Hua, Z., Zhang, R., Liu, G., Wen, W. et al. (2020). Automated arrhythmia classification based on a combination network of CNN and LSTM. *Biomedical Signal Processing and Control*, 57: 101819.

Chen, H., Dou, Q., Yu, L., Qin, J., Heng, P. -A. et al. (2018). Voxresnet: Deep voxel wise residual networks for brain segmentation from 3d MR images. *NeuroImage*, 170: 446–455.

Chen, J., Yang, L., Zhang, Y., Alber, M., Chen, D. Z. et al. (2016). Combining fully convolutional and recurrent neural networks for 3d biomedical image segmentation. *In Advances in Neural Information Processing Systems*, pp. 3036–3044.

Christ, P. F., Elshaer, M. E. A., Ettlinger, F., Tatavarty, S., Bickel, M. et al. (2016). Automatic liver and lesion segmentation in CT using cascaded fully convolutional neural networks and 3d conditional random fields. In *International Conference on Medical Image Computing and Computer-Assisted Intervention*. Springer, pp. 415–423.

Çiçek, O., Abdulkadir, A., Lienkamp, S. S., Brox, T., Ronneberger, O. et al. (2016). 3d u-net: Learning dense volumetric segmentation from sparse annotation. In *International Conference on Medical Image Computing and Computer-Assisted Intervention*. Springer, pp. 424–432.

Çınar, A. and Tuncer, S. A. (2021). Classification of normal sinus rhythm, abnormal arrhythmia and congestive heart failure ECG signals using LSTM and hybrid CNN-SVM deep neural networks. *Computer Methods in Biomechanics and Biomedical Engineering*, 24(2): 203–214.

Codella, N. C., Gutman, D., Celebi, M. E., Helba, B., Marchetti, M. A. et al. (2018). Skin lesion analysis toward melanoma detection: A challenge at the 2017 International Symposium on Biomedical Imaging (ISBI), hosted by the International Skin Imaging Collaboration (ISIC). In *IEEE 15th International Symposium on Biomedical Imaging (ISBI 2018)*, pp. 168–172.

Commandeur, F., Goeller, M., Betancur, J., Cadet, S., Doris, M. et al. (2018). Deep learning for quantification of epicardial and thoracic adipose tissue from non-contrast ct. *IEEE Transactions on Medical Imaging*, 37(8): 1835–1846.

Currie, G. M. (2019). Intelligent imaging: Artificial intelligence augmented nuclear medicine. *Journal of Nuclear Medicine Technology*, 47(3): 217–222.

Currie, G., Hawk, K. E., Rohren, E., Vial, A., Klein, R. et al. (2019). Machine learning and deep learning in medical imaging: Intelligent Imaging. *Journal of Medical Imaging and Radiation Sciences*, 50(4): 477–487.

Deep Learning In Health Care—A Ray of Hope in the Medical World, Homepage https://www.analyticsvidhya.com/blog/2021/08/deep-learning-in-health-care-a-ray-of-hope-in-the-complex-medical-world last accessed on 12/09/2021.

Deep Learning Market—Growth, Trends, Forecasts (2020–2025), Homepage https://www.mordorintelligence.com/industry-reports/deep-learning last accessed on 12/09/2021.

Deep Learning, Homepage https://www.ibm.com/cloud/learn/deep-learning last accessed on 12/09/2021.

Dou, Q., Yu, L., Chen, H., Jin, Y., Yang, X. et al. (2017). 3D deeply supervised network for automated segmentation of volumetric medical images. *Medical Image Analysis*, 41: 40–54.

Eltrass, A. S., Tayel, M. B. and Ammar, A. I. (2021). A new automated CNN deep learning approach for identification of ECG congestive heart failure and arrhythmia using constant-Q non-stationary Gabor transform. *Biomedical Signal Processing and Control*, 65: 102326.

Erdenebayar, U., Kim, Y. J., Park, J. -U., Joo, E. Y., Lee, K. -J. et al. (2019). Deep learning approaches for automatic detection of sleep apnea events from an electrocardiogram. *Computer Methods and Programs in Biomedicine*, 180: 105001.

Fakoor, R., Ladhak, F., Nazi, A. and Huber, M. (2013). Using deep learning to enhance cancer diagnosis and classification. In *Proc. of the International Conference on Machine Learning*, vol. 28. ACM, New York, USA, pp. 3937–3949.

Feng, X., Yang, J., Laine, A. F. and Angelini, E. D. (2017). Discriminative localization in CNNs for weakly-supervised segmentation of pulmonary nodules. In *International Conference on Medical Image Computing and Computer-Assisted Intervention*. Springer, pp. 568–576.

Fujita, H. and Cimr, D. (2019). Computer aided detection for fibrillations and flutters using deep convolutional neural network. *Information Sciences*, 486: 231–239.

Gibson, E., Robu, M. R., Thompson, S., Edwards, P. E., Schneider, C. et al. (2017). Deep residual networks for automatic segmentation of laparoscopic videos of the liver. In *Medical Imaging:*

Image-Guided Procedures, Robotic Interventions, and Modeling, vol. 10135. International Society for Optics and Photonics, 2017.

Golan, R., Jacob, C. and Denzinger, J. (2016). Lung nodule detection in ct images using deep convolutional neural networks. In *IEEE International Joint Conference on Neural Networks (IJCNN)*, pp. 243–250.

He, K., Zhang, X., Ren, S. and Sun, J. (2016). Deep residual learning for image recognition. In *Proc. of the IEEE Conference on Computer Vision and Pattern Recognition*, pp. 770–778.

He, K., Zhang, X., Ren, S. and Sun, J. (2016). Identity mappings in deep residual networks. In *European Conference on Computer Vision*. Springer, pp. 630–645.

Hochreiter, S. and Schmidhuber, J. (1997). Long short-term memory. *Neural Computation*, 9(8): 1735–1780.

Hu, B., Wei, S., Wei, D., Zhao, L., Zhu, G. et al. (2019). Multiple time scales analysis for identifying congestive heart failure based on heart rate variability. *IEEE Access*, 7: 17862–17871.

Huang, J., Chen, B., Yao, B. and He, W. (2019). ECG arrhythmia classification using STFT-based spectrogram and convolutional neural network. *IEEE Access*, 7: 92871–92880.

Hwang, S. and Kim, H. -E. (2016). Self-transfer learning for weakly supervised lesion localization. In *International Conference on Medical Image Computing and Computer-assisted Intervention*. Springer, pp. 239–246.

Ioffe, S. and Szegedy, C. (2015). Batch normalization: Accelerating deep network training by reducing internal covariate shift. In *International Conference on Machine Learning. PMLR*, pp. 448–456.

Isler, Y., Narin, A., Ozer, M. and Perc, M. (2019). Multi-stage classification of congestive heart failure based on short-term heart rate variability. *Chaos, Solitons & Fractals*, 118: 145–151.

Izci, E., Ozdemir, M. A., Degirmenci, M. and Akan, A. (2019). Cardiac arrhythmia detection from 2D ECG images by using deep learning technique. In *2019 Medical Technologies Congress (TIPTEKNO)*. IEEE, pp. 1–4.

Kamnitsas, K., Chen, L., Ledig, C., Rueckert, D., Glocker, B. et al. (2015). Multi-scale 3D convolutional neural networks for lesion segmentation in brain MRI. *Ischemic Stroke Lesion Segmentation*, 13: 46.

Kamnitsas, K., Ledig, C., Newcombe, V. F., Simpson, J. P., Kane, A. D. et al. (2017). Efficient multi-scale 3d cnn with fully connected crf for accurate brain lesion segmentation. *Medical Image Analysis*, 36: 61–78.

Kawahara, J., BenTaieb, A. and Hamarneh, G. (2016). Deep features to classify skin lesions. In *IEEE 13th International Symposium on Biomedical Imaging (ISBI)*, pp. 1397–1400.

Kleesiek, J., Biller, A., Urban, G., Kothe, U., Bendszus, M. et al. (2014). Ilastik for multimodal brain tumor segmentation. *Proc. MICCAI BraTS (Brain Tumor Segmentation Challenge)*, pp. 12–17.

Kronman, A. and Joskowicz, L. (2016). A geometric method for the detection and correction of segmentation leaks of anatomical structures in volumetric medical images. *International Journal of Computer Assisted Radiology and Surgery*, 11(3): 369–380.

Litjens, G., Kooi, T., Bejnordi, B. E., Setio, A. A. A., Ciompi, F. et al. (2017). A survey on deep learning in medical image analysis. *Medical Image Analysis*, 42: 60–88.

Long, J., Shelhamer, E. and Darrell, T. (2015). Fully convolutional networks for semantic segmentation. In *Proc. of the IEEE Conference on Computer Vision and Pattern Recognition*, pp. 3431–3440.

Mahmud, T., Hossain, A. R. and Fattah, S. A. (2019). EcgDeepNet: A deep learning approach for classifying ECG beats. In *2019 7th International Conference on Robot Intelligence Technology and Applications (RiTA)*. IEEE, pp. 32–37.

Moeskops, P., Wolterink, J. M., van der Velden, B. H., Gilhuijs, K. G., Leiner, T. et al. (2016). Deep learning for multi-task medical image segmentation in multiple modalities. In *International Conference on Medical Image Computing and Computer-Assisted Intervention*. Springer, pp. 478–486.

Nie, D., Wang, L., Gao, Y. and Shen, D. (2016). Fully convolutional networks for multimodality isointense infant brain image segmentation. In *IEEE 13th International Symposium on Biomedical Imaging (ISBI)*, pp. 1342–1345.

Panda, R., Jain, S., Tripathy, R. and Acharya, U. R. (2020). Detection of shockable ventricular cardiac arrhythmias from ECG signals using FFREWT filter-bank and deep convolutional neural network. *Computers in Biology and Medicine*, 124: 103939.

Pandey, S. K. and Janghel, R. R. (2019). Automatic detection of arrhythmia from imbalanced ECG database using CNN model with SMOTE. *Australasian Physical & Engineering Sciences in Medicine*, 42(4): 1129–1139.

Petmezas, G., Haris, K., Stefanopoulos, L., Kilintzis, V., Tzavelis, A. et al. (2021). Automated atrial fibrillation detection using a hybrid CNN-LSTM network on imbalanced ECG datasets. *Biomedical Signal Processing and Control*, 63: 102194.

Prasoon, A., Petersen, K., Igel, C., Lauze, F., Dam, E. et al. (2013). Deep feature learning for knee cartilage segmentation using a triplanar convolutional neural network. In *International Conference on Medical Image Computing and Computer-Assisted Intervention*. Springer, pp. 246–253.

Ronneberger, O., Fischer, P. and Brox, T. (2015). U-net: Convolutional networks for biomedical image segmentation. In *International Conference on Medical Image Computing and Computer-Assisted Intervention*. Springer, pp. 234–241.

Roth, H. R., Lu, L., Seff, A., Cherry, K. M. Hoffman, J. et al. (2014). A new 2.5 d representation for lymph node detection using random sets of deep convolutional neural network observations. In *International Conference on Medical Image Computing and Computer-Assisted Intervention*. Springer, pp. 520–527.

Shen, D., Wu, G. and Suk, H. -I. (2017). Deep learning in medical image analysis. *Annual Review of Biomedical Engineering*, 19: 221–248.

Simonyan, K. and Zisserman, A. (2014). Very deep convolutional networks for large-scale image recognition. *arXiv preprint arXiv:1409.1556*.

Srivastava, N., Hinton, G., Krizhevsky, A., Sutskever, I., Salakhutdinov, R. et al. (2014). Dropout: A simple way to prevent neural networks from overfitting. *The Journal of Machine Learning Research*, 15(1): 1929–1958.

Srivastava, N., Mansimov, E. and Salakhudinov, R. (2015). Unsupervised learning of video representations using LSTMs. In *International Conference on Machine Learning*. PMLR, pp. 843–852.

The Evolution of Deep Learning Homepage https://towardsdatascience.com/the-deep-history-of-deep-learning-3bebeb810fb2 last accessed on 12/09/2021.

Wang, L., Zhou, W., Chang, Q., Chen, J., Zhou, X. et al. (2019). Deep ensemble detection of congestive heart failure using short-term RR intervals. *IEEE Access*, 7: 69559–69574.

Xingjian, S., Chen, Z., Wang, H., Yeung, D. -Y. Wong, W. -K. et al. (2015). Convolutional LSTM network: A machine learning approach for precipitation now casting. In *Advances in Neural Information Processing Systems*, pp. 802–810.

Xu, X. and Liu, H. (2020). ECG heartbeat classification using convolutional neural networks. *IEEE Access*, 8: 8614–8619.

Yu, L., Chen, H., Dou, Q., Qin, J. and Heng, P. -A. (2016). Automated melanoma recognition in dermoscopy images via very deep residual networks. *IEEE Transactions on Medical Imaging*, 36(4): 994–1004.

Zagoruyko, S. and Komodakis, N. (2016). Wide residual networks. *arXiv preprint arXiv:1605.07146*.

Zeiler, M. D. and Fergus, R. (2014). Visualizing and understanding convolutional networks. In *European Conference on Computer Vision*. Springer, pp. 818–833.

Zeng, G. and Zheng, G. (2018). Multi-stream 3d fcn with multi-scale deep supervision for multimodality isointense infant brain mr image segmentation. In *IEEE 15th International Symposium on Biomedical Imaging (ISBI 2018)*, pp. 136–140.

Zhang, W., Li, R., Deng, H., Wang, L., Lin, W. et al. (2015). Deep convolutional neural networks for multimodality iso intense infant brain image segmentation. *Neuro Image*, 108: 214–224.

Zhou, S. and Tan, B. (2020). Electrocardiogram soft computing using hybrid deep learning CNN-ELM. *Applied Soft Computing*, 86: 105778.

Chapter 7

Automatic Classification of fMRI Signals from Behavioral, Cognitive and Affective Tasks Using Deep Learning

Cemre Candemir,[a,*] *Osman Tayfun Bişkin,*[b] *Mustafa Alper Selver*[c] and *Ali Saffet Gönül*[d]

ABSTRACT

Functional magnetic resonance imaging (fMRI) studies have achievements towards boosted promising outcomes due to deep applications. One of the emerging but rarely studied fields in this area is the decoding of differently stimulated tasks. Regarding this, a two-stage fMRI signal classification system is proposed using deep neural network (DNNs). In this respect, a unique collection of fMRI data sets consisting of behavioural, cognitive, and affective task-based (i.e., motor, memory, and emotion tasks) signals were gathered together with resting data from 58 volunteer participants. In the first stage of the proposed

---

[a] Research Assistant, International Computer Institute, Ege University, Izmir, TURKEY.
[b] Department of Electrical-Electronics Engineering, Burdur Mehmet Akif Ersoy University, Burdur, TURKEY.
[c] Department of Electrical and Electronics Engineering, Dokuz Eylül University, Izmir, TURKEY.
[d] Department of Psychiatry, SoCAT LAB, Faculty of Medicine, Ege University, Izmir, TURKEY.
* Corresponding author: cemre.candemir@ege.edu.tr

system, it is aimed to accurately determine the task classes to which a given signal belongs. In the second stage, the goal is to determine which phase of the fMRI experiment a given signal part belongs to. The Long-Short Term Memory (LSTM) and Gates Recurrent Units (GRU) approaches are chosen and their result are compared in terms of precision, recall, F1 score, and computation time. Extensive simulations are carried out and the results show that LSTM outperforms the GRU model for the classification of emotion and memory phases of participants. However, GRU is observed to be slightly better than the LSTM model for the task classification process. The achieved results also have higher accuracy values when compared to similar studies in the literature.

1 Introduction

Understanding the structure and working mechanism of the human brain has been one of the most intriguing topics for quite a long time. Although the *in vivo* brain anatomy has been largely discovered with the help of various medical imaging methods, understanding its working mechanism and neural connections has gained considerable speed in the last few decades thanks to functional imaging properties.

Functional magnetic resonance imaging (fMRI) is a non-invasive medical imaging technique used in both routine clinical workflow and research studies. It allows imaging of active regions in the brain that react to a given stimulus. One of the advantages of fMRI is that it does not use radiation unlike x-ray or positron emission tomography (PET). Thus, it provides a safe methodology. Moreover, fMRI is one of the medical imaging methods, which provides sufficiently high spatial resolution. It can be used for a great variety of research questions and many different tasks can be developed related to them such as memory, speaking, visual information, movement, language, emotions, and psychiatric disorders. Even if the stimulus comes from either different kinds of tasks, associated regions can be inferred. With these advantages and capabilities, fMRI has been a groundbreaking method in understanding the working mechanism of the brain.

The acquisition of the brain signals through fMRI relies on the theory that convolves given stimuli with the hemodynamic response function (HRF). According to the underlying fact that active brain regions need more oxygen, activity patterns in those regions can be determined from blood-oxygen-level dependent (BOLD) signals that respond to the various stimuli. Even though the acquisition of signals is still the same in theory, fMRI analysis has changed dramatically in the last decade. The conventional approach aims to map given stimuli with the regions while taking into account the activity of each brain voxel (Monti, 2011). In parallel with the development of the processing and analyzing environments, this common approach has provided tremendous perspectives and advances about the neurological working principles of the brain (Igelström and Graziano, 2017). Moreover, it leads up to some impressive studies which shows that it is possible to predict and detect the person's behavior or decisions using neural signals precisely. Several interesting researches through such analysis reveal the automatic detection in emotion changes (Candemir et al., 2021), prediction of the purchase decisions of individuals (Knutson et al., 2007), forecast of the aggregate choice on stock price dynamics (Stallen et al., 2021), predictions of the votes in elections (Knutson et al., 2006) (Rule et al., 2010), and intentions before actions (Gallivan et al., 2011). Here, it is worth pointing out that activation regions could only be discovered where the given stimulus set is known.

On the other hand, it has been shown that a reverse approach is possible with the acceleration of machine learning methods, especially in the last decade. This inverse approach, called brain decoding or neural decoding, has sparked great excitement in understanding the brain. The underlying idea is that brain activity patterns gathered from BOLD signals can be used to predict distinctive tasks and/or cognitive states that respond to a variety of specific stimuli. Prior studies show that deep neural networks (DNNs) can be used as a powerful tool for modelling the BOLD signals. DNNs have been successfully trained to decode cognitive tasks such as auditory stimuli (Zhao et al., 2018), speaking (Correia et al., 2014) and motor imagery (Pilgramm et al., 2015) in previous studies. Moreover, some studies show that it is possible to reconstruct the visual object stimulus from the fMRI activities of the individuals by combining DNNs with fMRI signals (Miyawaki et al., 2008; Shen et al., 2019). Reconstructing colored face images from fMRI activity is also reported with moderate accuracy (Cowen et al., 2014). Nevertheless, the wide range of existing studies have been designed to decode single activity. Classification of different cognitive tasks conducted with multi subjects are rarely reported in this area (Gao et al., 2019; Onal et al., 2017; Richiardi et al., 2011).

In this respect, the contribution of this study has two folds. First, a new data set is constructed for the multi-activity classification task and second, an effective strategy is developed for obtaining higher accuracy compared to the literature. Unfortunately, signal classification from different tasks is a very challenging goal in several aspects. The first difficulty is individual differences and multi-subject variability. These make the classification and reconstruction period much more complicated, which is the most sophisticated issue in the area. Second, the low signal-to-noise ratio (SNR) of the high dimensional fMRI signals make the classification task much more sophisticated. Here, it is critical to remove or suppress the noise as much as possible while keeping the valuable signals in. And the third question is on how the activity patterns in noisy brain signals can be efficiently represented through the machine

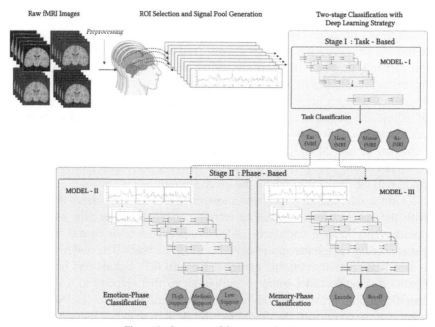

Figure 1: Summary of the proposed system.

learning based classifiers. We believe that this third issue is especially crucial to understand the complex relationships of cognitive mechanisms using machine learning methods.

A classifier is presented to decode three different type of tasks (i.e., cognitive, behavioral, and affective) using long short-term memory (LSTM) and Gates Recurrent Units (GRU) trained with the fMRI signals. In literature, both models are intensively used in time series data which include temporal dependency. As claimed in (Li and Fan, 2018) that, LSTM is a powerful tool for brain decoding tasks. Although GRU seems similar to LSTM models in many aspects, it was proposed to simplify the complexity of LSTM models.

During the design of the system, a data set, which is a collection of fMRI tasks acquired from multi subjects, is used. The whole data set contains four different fMRI subsets conducted with the following tasks: (1) resting-state fMRI (rs-fMRI), (2) motor task (motor-fMRI), (3) emotion task (Em-fMRI) and (4) memory task (mem-fMRI). Here, rs-fMRI presents spontaneous neuronal activity. Motor-fMRI is the behavioral task that exhibits the neuronal activity while performing a motor action. Em-fMRI is the affective task, and it presents the emotional activity while the subject is stimulated with the intentional emotional change phases. Mem-fMRI is the working visual memory task, i.e., the cognitive task, that consists of encoding, decoding, and resting blocks. After the region of interests (ROIs) are determined for each task with group analysis, BOLD signals are extracted from the voxels located in each ROI to be fed into the LSTM. The flowchart of the proposed system is given in Fig. 1. In our study, to the best of our knowledge, GRU is used for the first time for classification on such a data set.

The main motivation of the study has two-folds. The first one is to accurately predict the type of stimulus from the activation patterns of BOLD signals in cases where the type of stimulus given, or the design of the task is unknown. The second is to determine which phase of the fMRI task a given piece belongs to. For these manners, a two-stage classification strategy is developed. Analysis showed that even if the type of task is unknown, classification of the tasks can be done with a high accuracy. Besides, the sub-phase classifications also achieved above average accuracy. Decoding the different cognitive tasks from fMRI signals is a remarkable and promising area to decipher the cognitive mechanisms with the help of deep learning methodologies and depending on the latest knowledge, such studies have been rarely reported.

The rest of the chapter is organized as follows: fMRI Acquisitions, fMRI tasks and associated data sets are introduced in Section 2. The proposed deep classification methodology is presented in detail in Section 3. Computational results and analysis are given in Section 4 and conclusions are presented in Section 5.

2 fMRI Acquisitions and Data Set Properties

In this study, a collection of data sets including different fMRI tasks is presented for the first time. The cognitive and affective task designs are unique in the literature. Apart from this, the behavioral and resting fMRI data were also acquired with the same scanner and calibration settings having approximate average age and inclusion criterion with the other fMRI data sets.

All fMRI tasks explained in this study were conducted by SoCAT[1] Research Lab in Ege University, Turkey. Functional images were acquired by a 3 Tesla (3T) Siemens whole-body MRI scanner. Parameters of the imaging procedures are flip angle (FA) = 90° and bandwidth = 2232 Hz/pixel for all scans. Repetition times (TR) and echo times (TE), the field of views (FOV), and slice numbers vary according to the tasks. The inclusion criteria of the subjects were their age (18–25 years old, mean age 22,12 ± 1.8), being healthy (i.e., not having a present or past mental illness, mental trauma, or unstable medical disease such as diabetes, and tensions), and being right-handed. In total, 58 subjects have taken part in the scans and all of them were volunteer university students. All conducted procedures were approved by the research ethics committee of the university.

---

[1] Standardization of Computational Anatomy Techniques, http://socatlab.com.

2.1 Data Set Descriptions

2.1.1 Resting State fMRI (rs-fMRI)

Unlike task-based fMRI models, resting state fMRI (rs-fMRI) focuses on spontaneous and intrinsically generated neuronal activity in BOLD signals. During the resting position (generally 6–10 minutes), participants are not stimulated by any task involving cognitive or motor stimuli. During the scan, the participants are asked to be comfortable, lie still and not to think about anything. The eyes of the participants may be open or closed. The regions active in the brain in the resting state are called the default mode networks and no activation is expected in areas outside of this network. rs-fMRI tasks last for 9 minutes. The entire brain imaging data consists of 37 slices, 64×64 matrix, FOV = 192×192 mm, voxel size $3 \times 3 \times 3$ mm, slice thickness 3 mm, TE = 30 ms, TR = 3000 ms. 180 image series are acquired for each subject in the data set.

2.1.2 Emotion fMRI (Em-fMRI)

The emotion-based affective Em-fMRI data set is obtained through a social support fMRI task. The task consists of a game aimed at triggering the alteration of emotional states through different levels of support. During the task, the participant and his three friends play a guessing game against a rival and earn some money at the end. The participant is the person who has had an MRI scan and actively plays the game. Three friends of the participant are in the role of the referee, who will distribute 10 points between their friends and the rival in the game. During the game (i.e., fMRI scan) referees are located in separate rooms and are not involved in the game actually. However, the participant and the referees are not informed of this. The whole task consists of three stages with different support levels: High support, medium support, and low support. The same 7-step procedure is followed for all support levels. The first step is the baseline screen, which is shown as a fixation cross (3s). The second step is the game screen. Here, a picture is shown to participants, which consists of a varying number of geometrical shapes (squares, triangles, stars, circles, etc.) and they are asked to predict how many squares are there in the picture (3s). As the third step, participants are asked to choose the suitable answer to the picture they saw on the previous screen with the buttons in their hands (3s). The possible options are >30, <30; >40, <40; >50, <50. Here, participants are also informed that if they missed the time or did not make any choice, an answer will be chosen randomly by the computer. Through this information, it was ensured that the subjects participated in the game better. In the fourth screen, there is an hourglass that exhibits the referees distributing points (6s). After that in the fifth screen, the participants see the both their own and their rival's questions and answers (3s). In the sixth and seventh screens, they see the point distribution (3s) and the amount of money they earned (3s) respectively.

Each 7-step procedure repeats 25 times in every stage (75 times in total). The first stage of the fMRI task is the high-support stage and participants win 80% of the game by feeling the support of their friends. However, the last part is the low-support stage and participants lose 80% by not feeling any support from their friends. It should be noted that the competitor's friends have no influence on these win-and-lose situations. Here, the task is set on triggering this emotional change of the person.

More details of the Em-fMRI task were presented in our previous study (Candemir et al., 2021). For each voxel, 600 image series are acquired with 37 slices, 64 × 64 matrix, 3.5 mm slice thickness (with 1 mm gap), 200 × 200 mm FOV. The voxel size is 3 × 3 × 3 mm, TE = 30 ms, TR = 3000 ms. The task lasted about 30 min (1818 s).

2.1.3 Motor fMRI

Motor fMRI is a behavioral task and during the fMRI acquisition, subjects are asked to do a series of motor stimulations. This is a finger-tapping experiment to identify motor-related regions in the brain. The motor-fMRI task has a block design that includes rest and movement parts repetitively. In the rest period, subjects stand without any movement for 60 s and thereafter, subjects are asked to perform repetitive finger opening and closing movements for 60 s during the movement period. Each of the rest and movement periods is repeated 10 times. The tapping is self-paced. In block design experiments, stimuli are given at fixed intervals. For this fMRI assignment, stimuli are given in terms of TR at $t(i)$ = {30, 90, 150, 210, 270, 330, 390, 450, 510, 570} for 30 TR. All tasks lasted 1200 seconds. A series of 600 images of each voxel are acquired with 23 slices with a matrix of size 64 × 64, voxel size is 3 × 3 × 3 mm, TR = 2000 ms, TE = 30 ms, FOV = 200 × 200 mm; 3.5 mm slice thickness (with 1 mm gap).

2.1.4 Memory fMRI (Mem-fMRI)

Memory fMRI is a cognitive fMRI task and has a block design that consists of the following sequential phases:

(1) Resting phase: During the first block of resting, subjects lie still for 300 seconds.

(2) Encoding phase: The encoding block comes after the resting block which is a set of neutral faces with their names shown to the subjects. The first screen

of the encoding phase is the baseline where subjects see a fixation cross for 30 s. Then in the second screen, a neutral face is shown with its name for 6 seconds and a one-second black screen follows. Here, subjects are asked to record the names associated with the faces. Face and black screen phases repeat six-times and the encoding block repeats four times. In total, subjects show 24 different faces during the encoding step. All encoding blocks last 288 seconds.

(3) Resting phase: The subjects lie still for 300 seconds.

(4) Recall phase: The recall phase consists of two main sub-phases: Face-recall and name-recall. During the face-recall phase, subjects decide whether they are familiar with the faces on the screen in 288 seconds. Face-recall starts with the baseline screen for 30 seconds, and after the neutral face appears for 6 seconds it is followed by a black screen for one second. Face and black screen periods repeat six-times and the face-recall phase repeats four times. Here, the subjects show 24 faces during the face-recall step. During the name-recall phase, subjects are asked to identify the names of the faces on the screen in 288 seconds. Name-recall starts with the baseline screen for 30 seconds, and after a neutral face appears for 6 seconds it is followed by a black screen for one second. Face and black screen periods repeat six-times and the name-recall phase repeats four times. Here, subjects show 24 face-name pairs during the name-recall step.

(5) Resting phase: Subjects lie still for 300 seconds.

The face data set that is used in the memory task is a subset of the FACES data set (Ebner et al., 2010). The original FACES data set contains images of naturalistic faces of 171 women and men expressing each of the six facial expressions: neutrality, sadness, disgust, fear, anger, and happiness. Only the faces which have neutral facial expressions are used in the memory task. Whole brain imaging data consists of 37 slices, 64 × 64 matrix, FOV = 192 × 192 mm, voxel size 3 × 3 × 3 mm, slice thickness 3 mm, TE = 30 ms, TR = 3000 ms.

300 and 484 volumes are acquired for each subject in resting states and encoding/recall phases respectively.

2.2 Preprocessing of Data Sets

MRI data requires a set of processing stages which consists of several steps. The main objectives during preprocessing are to align the images to a standard space and to remove the noise as much as possible so that the validity of the group analysis improves while the physiologic-related artifacts, such as breathing and the heart rate and scanner-related artifacts, such as system noise and radio frequency artifacts, are minimized.

The preprocessing of the functional data was done with the Statistical Parametric Maps (SPM) version 12.[2] All functional images were corrected for involuntary head motion, which is known as the realignment step, and afterward, the slices were synchronized temporally in the slice timing step. Later, the structural scans of the subjects were registered to the mean images of fMRI scans, which is referred to as co-registration. The next preprocessing step was segmentation where the brain is separated from its surrounding tissues. During segmentation, the structural image was also normalized to a global standard space, which is the standard Montreal Neurological Institute (MNI) single-subject template. Finally, normalized images were spatially smoothed with an 8 mm isotropic Gaussian kernel.

2.3 ROI Selection and Signal Pool Generation

To determine the regions of interest (ROIs), a two-level analysis should be performed with the fMRI data first. The first-level analysis shows the activation maps at the individual level, whereas the second-level analysis exhibits the active areas in common at the group level. Once the second-level analysis is completed, the ROI selection could be done. All first and second-level analyses have been run with the SPM12 for each data set.

Since the three stimuli-induced tasks come from different behavioral tasks and hypothesize different areas, three ROIs are determined for each one: Nucleus Accumbens (NAcc) for the emotion task, Broadmann Area-4 (BA4) for the motor task, and finally Occipital Face Area (OFA) for the memory task, as shown in Fig. 2. NAcc, depicted in red in Fig. 2, plays an important

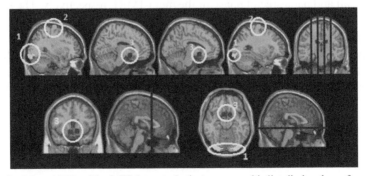

Figure 2: Selected ROIs of the fMRI data sets. In the top row, multi-slice display shows from left to right OFA (1) and BA4 (2), Nacc (3) is seen for both left and right hemispheres in the sagittal plane. In the lower-left corner, left and right NAcc is presented in the coronal plane. In the lower right corner, both Nacc and OFA can be seen in the axial plane.

---

[2] http://www.fil.ion.ucl.ac.uk/spm/.

role in reward-related behaviors, whereas BA4, shown in cyan, is associated with the motor (such as the finger, hand, wrist movements and tongue, face, and mouth movement) and somatosensory functions (such as perception of limb movements, attention to action). OFA, shown in yellow, is responsible for face processing, responses to the different face stimuli, and recognizing familiar faces. All ROIs are presented in Fig. 2 in sagittal, coronal, and axial axes, respectively.

All regions, determined as active have a cluster size >10, adjacent voxels with a global maxima that meet the Family Wise Error (FWE)-corrected threshold of $p < 0.05$. After the voxel coordinates are identified with SPM12, ROI masks are created manually by WFU Pick Atlas.[3] For the rs-fMRI, the signals from each ROI (i.e., NAcc, BA4, and OFA) are obtained separately.

An fMRI signal (i.e., BOLD signal) ($x(t)$) is expressed mathematically as the convolution of the stimulus ($s(t)$) with the hemodynamic response function (HRF) ($h(t)$) with $x(t) = s(t) * h(t)$. The signal of a voxel at the time is represented by $y_i(t) = \beta_i * x(t) + \epsilon_i(t)$, where β_i are the weights of each regressor x (i.e., given neural inputs) and $\epsilon_i(t)$ are the noise-induced errors ($\epsilon \stackrel{iid}{\sim} N(0, \sigma^2)$) (Kiebel et al., 2007). A "BOLD" signal could be modelled for each, and every voxel as exhibited in Fig. 3.

The total number of BOLD signals that comprise the signal pool is 24,674 for 58 subjects. They were obtained for each task (i.e., Em-fMRI, motor-fMRI, and mem-fMRI) and rs-fMRI scan. As above-mentioned, all the ROIs were also used to get signals from rs-fMRI. The signals are of size 600 × 50 for motor-fMRI, 180 × 50 for motor-resting, 600 × 102 for Em-fMRI, 180 × 102 for Em-resting, 180 × 90 for each 1st, 2nd and 3rd resting phases of mem-fMRI, 100 × 90 for the encoding phase, 192 × 76 and 192 × 93 for the face-recall and name-recall phase of the mem-fMRI, respectively. The signals of all participants are obtained through the same procedure and have the same vectorial size for all. Examples of the signals that constitutes the signal pool are given in Fig. 3b. It should be note that the signals are under the effect of both stimuli type and inter-subject differences. All signals are de-trended and normalized to zero mean before being fed into the proposed model.

3 Proposed Method

3.1 Problem Statement

In this study, we are concerned with the classification of fMRI signals employing deep learning-based methods. Classification of signals is a process

---
[3] https://www.nitrc.org/projects/wfu_pickatlas/.

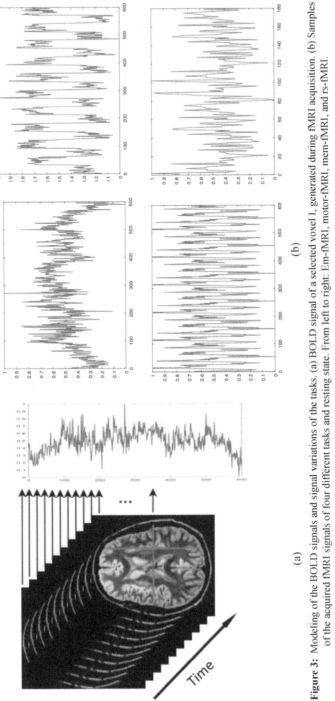

Figure 3: Modeling of the BOLD signals and signal variations of the tasks. (a) BOLD signal of a selected voxel I, generated during fMRI acquisition. (b) Samples of the acquired fMRI signals of four different tasks and resting state. From left to right: Em-fMRI, motor-fMRI, mem-fMRI, and rs-fMRI.

to categorize a related signal into sub-categories by using some inherent features of the dataset. Let a sequence $\mathbf{x} = [x_1 \ldots x_T]$ denote time series data with $x_t \in \mathbb{R}^d$ where d represents the number of dimensions of the time-series data, x_t, at time instants t. The aim of the classification problem is finding a nonlinear mapping function, $f(.)$, which matches a sequence with a predefined labeled class. Here, we propose employing cascaded LSTM and GRU models for the classification of fMRI signals.

The flowchart of the fMRI signal classification system proposed in this study is given in Fig. 1 within the red-lined boxes. The proposed system consists of two main classification stages. Stage 1 is the task classification stage in which the resting, behavioral, cognitive, affective task classes (Em-fMRI, rs-fMRI, motor fMRI, and Mem-fMRI), is determined using the acquired fMRI signals. On the other hand, in Stage II of the proposed system, Em-fMRI and Mem-fMRI signals are classified into their sub-tasks given to a participant to perform during the experiments. Em-fMRI signal is classified into high support, medium support, and low support classes that reflect participants' emotional changes during the experiment. On the other hand, Mem-fMRI signal is classified into 2 groups, encoding and recall, indicating the main phases of the Mem-fMRI task.

The reason we propose a system with two stages is that fMRI signals acquired from participants do not allow us to classify the signal into its sub-tasks or phases using only one stage system. For instance, an Em-fMRI signal includes three phases: high support, medium support, and low support. In order to determine the phases of an Em-fMRI signal, we need to divide the signal into three equal parts. By doing this, DNNs can classify the phase of each part of the Em-fMRI signal. Therefore, in order to accomplish this one has to know which task classes the fMRI signal belong to. Then, before employing the DNNs to determine the signal phase, we apply proper process according to the signal task class. Thus, it is not possible to accomplish phase classification using only one stage because of the structure of our data acquired from the participant during the experiments.

3.2 LSTM Models

LSTM was originally proposed in (Hochreiter and Schmidhuber, 1997) to cope with the vanishing gradient problem which limits the capability of Recurrent Neural Networks used for time series applications having temporal dependencies (Hochreiter and Schmidhuber, 1997). Information flow throughout the LSTM memory is controlled by three gates, input, output and forget gates (Gers et al., 2000). Input and output gates control the flow of input and output activation information, respectively. In addition, the input

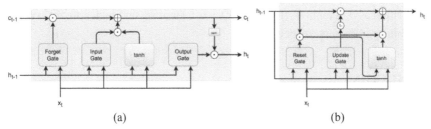

Figure 4: Architecture of (a) LSTM and (b) GRU units.

gate controls the cell state together with the forget gate (Zeroual et al., 2020). Forget gates reset the memory of the cell in case the cell memory is not needed anymore (Gers et al., 2000). The architecture of an LSTM unit is given in Fig. 4a. Let $\mathbf{x}_t$ be the input sequence at time t, then the input gate (i_t) is $i_t = \sigma(W_i[h_{t-1}, x_t] + b_i)$. In addition, output gate (o_t) and forget gate (f_t) are given as $o_t = \sigma(W_o[h_{t-1}, x_t] + b_o)$ and $f_t = \sigma(W_f[h_{t-1}, x_t] + b_f)$, respectively. Here W_i, W_o, and W_f represent input, output, and forget weight parameters, respectively. The parameters b_i, b_o, and b_f are the bias parameters and h_t is the hidden state vector, or sometimes called, output state, at time t. In addition, $\sigma(.)$ given in Fig. 4b is the sigmoid function. The hidden (h_t) and cell states (c_t) are given as $c_t = i_t \odot \tilde{c}_t + f_t \odot c_{t-1}$ and $h_t = o_t \odot \tanh(c_t)$, respectively. Here $\odot$ represents Hadamard product and $\tanh(.)$ is the hyperbolic tangent function. On the other hand, $\tilde{c}_t$ parameter in the cell state is given as $\tilde{c}_t = \tanh(W_c[h_{t-1}, x_t] + b_c)$. Here, W_c and b_c represent the weight and bias parameters, respectively.

3.3 GRU Models

GRU was introduced to simplify the complexity of the LSTM architecture by decreasing the number of parameters in LSTM models (Cho et al., 2014; Chung et al., 2014; Shahid et al., 2020; Zeroual et al., 2020). The GRU model has two gates: update and reset gates. The update gate presented in GRU architecture is a kind of combination of input gates and forget gates introduced in the LSTM model. The update gate in the model controls the memory which is transferred to the new state. The mission of the reset gate in GRU architecture is similar to the forget gate in LSTM model, and the role of this gate is to forget the past information which is not used anymore. Figure 4b shows the architecture of the GRU model. The update gate (u_t) and reset gate (r_t) in GRU are given by $u_t = \sigma(W_u[h_{t-1}, x_t] + b_u)$ and $r_t = \sigma(W_r[h_{t-1}, x_t] + b_r)$, respectively. Here, W_u and W_r denote update and reset weight parameters, respectively. On the other hand, b_u and b_r are the bias parameters. h_t is the hidden state (or output state) vector at the time t. We can give the hidden state at the time t as $h_t = u_t \odot \tilde{h}_t + (1 - u_t) \odot c_{t-1}$, where $\tilde{h}_t$ can be written as $\tilde{h}_t = \tanh(W_h[r_t \odot h_{t-1}, x_t] + b_h)$. Here W_h and b_h denote the weight and bias parameters, respectively.

4 Results

First of all, since the length of fMRI signals are different, they are padded before feeding the deep neural network at the beginning of Stage I. In this way, all fMRI signals acquired by different ROIs become 1 × 600 vectors, and therefore, each signal can be represented by $\mathbf{x} = [x_1 \ldots x_T]$ where $x_t \in \mathbb{R}$, and $t = 1,\ldots,T$ and $T = 600$. The vector $\mathbf{x}$ is the input signal of Model I given in Fig. 1.

In the experiments performed in Stage I, a two-layer network is used for Model I in order to classify the fMRI signal into Em-fMRI, rs-fMRI, motor fMRI, and Mem-fMRI tasks. In order to find the optimal model parameters for all numerical experiments employed in this study, we search the hidden units in {50, 75, 100, 125, 150, 175, 200, 250}, learning rate, lr, in $\{10^{-2}, 5 \times 10^{-3}, 10^{-3}, 5 \times 10^{-4}, 10^{-4}\}$, and number of epochs in {200, 300,...,1000} on the validation set. To train the deep neural network, the data set is divided into training and test data by considering real-life scenarios. Therefore, at each training process, fMRI signals acquired from one participant are used as test data while fMRI signals of the other participants are considered as training data. All simulations accomplished in this study are performed on a PC employing i7-9750H CPU with a memory of 16 GB and a processor speed of 2.6 GHz. To speed up the model training, algorithms are performed by utilizing the NVIDIA GTX 1660 Ti GPU with a memory of 6 GB.

In the first experiment, we employed LSTM deep learning models having two layers with 100 and 75 hidden units, respectively. The first and second hidden units are followed by dropout layers with 0.2 and 0.1 dropout ratios, respectively. For all networks, a cross-entropy function is the loss function which is used to estimate the loss of the model during the training process. Adam optimizer (Kingma and Ba, 2014) is utilized for minimizing the loss function during the training and the learning rate of the optimizer is taken as $lr = 0.001$. The batch size is chosen as 128 and models are trained for an epoch number, 300. Next, we employed GRU layers instead of LSTM layers in Model I by utilizing the same parameters. Numerical results for the task classification experiment are given in Table 1. Classification performances of models are computed in terms of precision, recall, and F_1 score metrics. In addition, the overall accuracy of the task classification experiment is presented in Table 3. We can see from Table 1 and Table 3 that GRU performs slightly better than the LSTM network in terms of all metrics for the classification

Table 1: Classification performances of models for behavioral, cognitive, and affective tasks classification experiment.

Task	LSTM			GRU		
	Precision	Recall	f1-score	Precision	Recall	f1-score
Emotion fMRI	98.02	96.98	97.49	99.33	97.22	98.27
Memory fMRI	93.09	88.74	90.87	94.97	88.70	91.73
Motor fMRI	95.92	94.99	94.95	80.36	90.00	84.91
rs-fMRI	95.41	97.26	96.33	95.46	98.07	96.75

of Em-fMRI, Mem-fMRI and rs-fMRI signals. However, LSTM outperforms the GRU model in the classification of motor fMRI signals.

In Stage II, we aim to determine the emotions of participants using their Em-fMRI signals and the memory activity of participants using their Mem-fMRI signals. For this purpose, Em-fMRI and Mem-fMRI signals are classified into their sub-tasks using deep learning-based methods given by Model II and Model III in Fig. 1, respectively. Model II is used to determine the emotion of a participant by classifying the Em-fMRI signals into high support, medium support, and low support classes. Let **x** be a 1 × 600 vector representing the Em-fMRI signal acquired from a voxel for a particular participant during the experiment. Then, it can be assumed that the Em-fMRI signal, **x**, includes all three emotions of the participant. In addition, we assume that the emotional changes occurring in the experiment have the same time intervals. Therefore, the Em-fMRI signal is divided into equal time intervals before feeding Model II. Let x_1, x_2, and x_3 represent the signal of emotion that arises in high support, medium support, and low support cases, respectively. Then, the x_1, x_2, and x_3 become 1 × 200 vectors obtained from the Em-fMRI signal, x. In order to classify the signal **x** into x_1, x_2, and x_3 classes, we employed LSTM and GRU models having two layers with 150 and 100 hidden units, respectively. For the LSTM network, the first and second hidden units are followed by dropout layers with 0.2 and 0.1 dropout ratios, respectively. On the other hand, the dropout ratio for both dropout layers are selected as 0.2 for GRU model. Optimizers of LSTM and GRU networks are Adam optimizers, and the learning rate of the optimizers is taken as 5×10^{-4}. For both models, the batch size and the epoch number are chosen as 2048 and 500, respectively. Classification performances of models in terms of precision, recall, and F_1 score metrics and overall accuracy are given in Table 2 and Table 3, respectively. In addition, emotion classification performances are plotted in Fig. 5 for each participant in the experiment. Figure 5a, Fig. 5c, and Fig. 5e show the results of classification employed by the LSTM model. On the other hand, performances of the GRU model are given in Fig. 5b, Fig. 5d, and Fig. 5f. Table 2, Table 3, and Fig. 5 indicate that,

Table 2: Classification performances of models for emotion and memory phase classification experiments.

		LSTM			GRU		
		Precision	Recall	F1-score	Precision	Recall	F1-score
Emotion	High Support	81.13	80.07	80.59	81.68	74.67	78.02
	Medium Support	76.23	70.75	73.39	72.35	74.18	73.26
	Low Support	79.52	86.27	82.76	80.20	85.05	82.55
Memory	Encoding Phase	99.06	98.60	98.83	98.33	97.89	98.07
	Recall Phase	99.25	99.50	99.38	98.93	99.13	99.02

Table 3: Overall accuracy of models for task classification, emotion classification, and memory phase classification experiments.

	LSTM			GRU		
	Task Classification	Emotion Classification	Memory Classification	Task Classification	Emotion Classification	Memory Classification
Accuracy	95.00	79.03	99.19	95.55	77.97	98.70

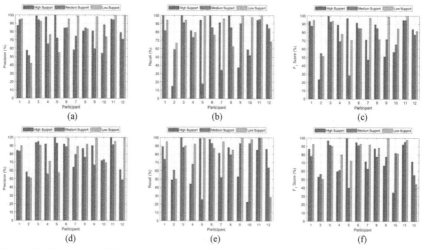

Figure 5: Emotion classification performances of models for each participant. The performance of LSTM network in terms of (a) precision, (b) recall, and (c) score and the performance of GRU network in terms of (d) precision, (e) recall, and (f) F_1 score.

LSTM outperforms the GRU network in terms of performance metrics for all phases of emotion task. Moreover, we can say that the overall accuracy of the LSTM model is also better than the GRU model.

In Stage II, we also determine the phases of the memory task utilizing the Mem-fMRI signals of participants. For this purpose, Model III given in

Fig. 1 is employed to classify the Mem-fMRI into two main phases: encoding and recall. The experiment in which the Mem-fMRI signal is acquired from participants is a block design fMRI task. Therefore, signals obtained at a specific time interval from a participant include either encoding or recall phases. The size of the acquired signal from a particular voxel is 1×192. Thus, 1×192 vector is given to Model III in order to determine the phases of the acquired signal. Similar to Model II, we employed LSTM and GRU models having two layers in Model III. The first and the second layers of Model III include 100 and 75 hidden layers, respectively. The optimizer for the networks is the Adam optimizer, and the learning rate of the optimizer is chosen as 5×10^{-4}. Batch size and epoch number are assigned to 512 and 300 for both LSTM and GRU models, respectively. The classification performance of models is given in Tables 2 and 3. Looking at Table 2 and Table 3, we can see that the LSTM performs slightly better than GRU models in terms of all performance metrics. In addition, Table 3 shows that the performance of models in the memory phase classification task outperforms the other tasks, emotion and task classification. Moreover, the performances of LSTM and GRU models for the memory phase classification are presented in Fig. 6 for each participant. Figure 7 shows the overall accuracy of emotion classification and memory phase classification processes employed by LSTM and GRU networks for each participant. We can see from Fig. 7 that, the LSTM model can classify the emotions of participants better than GRU model. On the other, the LSTM model outperforms the GRU model in the classification of memory phases.

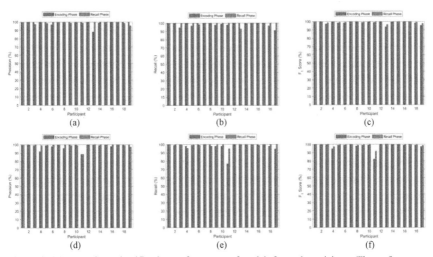

Figure 6: Memory phase classification performances of models for each participant. The performance of LSTM network in terms of (a) precision, (b) recall, and (c) score and the performance of GRU network in terms of (d) precision, (e) recall, and (f) F_1 score.

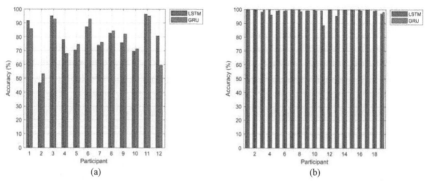

Figure 7: Overall accuracy of (a) emotion classification and (b) memory phase classification employing LSTM and GRU networks for each participant.

With the proposed method, obtained results have higher accuracy when compared to recent studies. Gao et al., reported cross-subject brain decoding results as 62.3%, 65.1%, and 67.5% for 8-layer, 16-layer and 32-layer LSTM models, respectively. Nevertheless, task-specific values are not available. Class-wise performances were obtained with ResNet-DTL method with following accuracy values: Emotion 77.9%, motor 82.5%, working memory 80.9% (Gao et al., 2019). In another study, the average of the classification performance of working memory varies between 66.56–90.7% accuracy in (Onal et al., 2017).

5 Conclusion

In this study, a classification strategy is proposed to identify behavioral, cognitive, and affective tasks, which constitute a challenging fMRI study. In this respect, a new and challenging data set is generated by collecting various fMRI studies. Thus, a collection of fMRI data sets, which consists of three task-based functional data and resting data, is used. Task-based fMRI data sets represent the emotion, motor, and memory actions, whereas the resting data represents the default mode. The properties of this data set are analyzed and reported in detail to initiate further and in-depth research at rarely studied multi-activity classification topics. Moreover, a deep learning-based two-stage classification strategy is also developed. In the first stage, an accurate prediction of the stimulus type is sought from the activation patterns. In the second stage, it is aimed to determine the sub-classes (i.e., phase of the task) from the given part of the signal. LSTM and GRU layers are used for the proposed system. Our simulation results show that the classification performance of the GRU network is slightly better than the LSTM network for the classification of emotion and memory actions. On the other hand, LSTM classifies the motor actions more precisely than the GRU model. In the second stage of the proposed classifier system, the LSTM model outperforms the GRU model in determining the phases of emotion and memory fMRI signals acquired from participants during the experiment. Overall, obtained results also have higher accuracy values when compared to similar studies.

Here, it is also worth mentioning the limitations of the study. The main limitation is the individual differences, which bring together multi-subject variability. This makes the classification process much complicated as it also creates significant differences in the amplitude of the BOLD signals. The other point is the high amount of noise of the BOLD signal, which must be removed or smoothed as much as possible. If this process is not done carefully, the information carried by the signal may be lost. The third limitation, like the other machine learning approaches, is the data set, and parameter dependency. Parameter adjustments are valid for this system and may require resetting at any change. The performance of the system also could possibly be enhanced by using larger data sets. However, it is a trade-off since increasing the number of data sets will also bring multi-subject variability together.

The detection of stimulus-related brain activations has also paved the way for various interesting applications not only in the medical field but also in interdisciplinary fields such as psychology (neuropsychology) and economics (neuroeconomics). In this way, it is possible to study understanding and modeling human decisions, choices, and behaviors. Our findings also give the idea to prefigure that the reverse approaches would take part in not only neuroimaging but also such interdisciplinary fields.

Acknowledgments

The authors have no conflicts of interest to declare.

References

Candemir, C., Gonul, A. S. and Selver, A. M. (2021). Automatic detection of emotional changes induced by social support loss using fMRI. *IEEE Transactions on Affective Computing*, 1–1. https://doi.org/10.1109/TAFFC.2021.3059965.

Cho, K., Van Merriënboer, B., Bahdanau, D. and Bengio, Y. (2014). On the properties of neural machine translation: Encoder-decoder approaches. *ArXiv Preprint ArXiv:1409.1259*.

Chung, J., Gulcehre, C., Cho, K. and Bengio, Y. (2014). Empirical evaluation of gated recurrent neural networks on sequence modeling. *ArXiv Preprint ArXiv:1412.3555*, 1–9.

Correia, J., Formisano, E., Valente, G., Hausfeld, L., Jansma, B. et al. (2014). Brain-based translation: FMRI decoding of spoken words in bilinguals reveals language-independent semantic representations in anterior temporal lobe. *Journal of Neuroscience*, 34(1): 332–338. https://doi.org/10.1523/JNEUROSCI.1302-13.2014.

Cowen, A. S., Chun, M. M. and Kuhl, B. A. (2014). Neural portraits of perception: Reconstructing face images from evoked brain activity. *NeuroImage*, 94: 12–22. https://doi.org/10.1016/j.neuroimage.2014.03.018.

Ebner, N. C., Riediger, M. and Lindenberger, U. (2010). FACES—A database of facial expressions in young, middle-aged, and older women and men: Development and validation. *Behavior Research Methods*, 42(1): 351–362. https://doi.org/10.3758/BRM.42.1.351.

Gallivan, J. P., McLean, D. A., Valyear, K. F., Pettypiece, C. E., Culham, J. C. et al. (2011). Decoding action intentions from preparatory brain activity in human parieto-frontal networks. *Journal of Neuroscience*, 31(26): 9599–9610. https://doi.org/10.1523/JNEUROSCI.0080-11.2011.

Gao, Y., Zhang, Y., Wang, H., Guo, X., Zhang, J. et al. (2019). Decoding behavior tasks from brain activity using deep transfer learning. *IEEE Access*, 7: 43222–43232. https://doi.org/10.1109/ACCESS.2019.2907040.

Gers, F. A., Schmidhuber, J. and Cummins, F. (2000). Learning to forget: Continual prediction with LSTM. *Neural Computation*, 12(10): 2451–2471.

Hochreiter, S. and Schmidhuber, J. (1997). Long short-term memory. *Neural Computation*, 9(8): 1735–1780. https://doi.org/10.1162/neco.1997.9.8.1735.

Igelström, K. M. and Graziano, M. S. A. (2017). The inferior parietal lobule and temporoparietal junction: A network perspective. *Neuropsychologia*, 105: 70–83. https://doi.org/10.1016/j.neuropsychologia.2017.01.001.

Kiebel, S. J., Kherif, A. P. and Holmes, C. (2007). CHAPTER 8—The General Linear Model. pp. 101–125. *In*: Friston, K., Ashburner, J., Kiebel, S., Nichols, T. and Penny, W. (eds.). *Statistical Parametric Mapping*. Academic Press. https://doi.org/10.1016/B978-012372560-8/50008-5.

Kingma, D. P. and Ba, J. L. (2014). Adam: A method for stochastic optimization. *ArXiv Preprint ArXiv:1412.6980*.

Knutson, B., Rick, S., Wimmer, G. E., Prelec, D., Loewenstein, G. et al. (2007). Neural predictors of purchases. *Neuron*, 53(1): 147–156. https://doi.org/10.1016/j.neuron.2006.11.010.

Knutson, K. M., Wood, J. N., Spampinato, M. V. and Grafman, J. (2006). Politics on the Brain: An fMRI Investigation. *Social Neuroscience*, 1(1): 25–40. https://doi.org/10.1080/17470910600670603.

Li, H. and Fan, Y. (2018). Brain decoding from functional MRI using long short-term memory recurrent neural networks. pp. 320–328. *In*: Frangi, A. F., Schnabel, J. A., Davatzikos, C., Alberola-López, C. and Fichtinger, G. (eds.). *Medical Image Computing and Computer Assisted Intervention—MICCAI 2018* (Vol. 11072). Springer International Publishing. https://doi.org/10.1007/978-3-030-00931-1_37.

Miyawaki, Y., Uchida, H., Yamashita, O., Sato, M., Morito, Y. et al. (2008). Visual image reconstruction from human brain activity using a combination of multiscale local image decoders. *Neuron*, 60(5): 915–929. https://doi.org/10.1016/j.neuron.2008.11.004.

Monti, M. M. (2011). Statistical analysis of fMRI time-series: A critical review of the GLM approach. *Frontiers in Human Neuroscience*, 5. https://doi.org/10.3389/fnhum.2011.00028.

Onal, I., Ozay, M., Mizrak, E., Oztekin, I. and Vural, F. T. Y. (2017). A new representation of fMRI signal by a set of local meshes for brain decoding. *IEEE Transactions on Signal and Information Processing over Networks*, 3(4): 683–694. https://doi.org/10.1109/TSIPN.2017.2679491.

Pilgramm, S., de Haas, B., Helm, F., Zentgraf, K., Stark, R. et al. (2015). Motor imagery of hand actions: Decoding the content of motor imagery from brain activity in frontal and parietal motor areas. *Human Brain Mapping*, 37(1): 81–93. https://doi.org/10.1002/hbm.23015.

Richiardi, J., Eryilmaz, H., Schwartz, S., Vuilleumier, P., Van De Ville, D. et al. (2011). Decoding brain states from fMRI connectivity graphs. *NeuroImage*, 56(2): 616–626. https://doi.org/10.1016/j.neuroimage.2010.05.081.

Rule, N. O., Freeman, J. B., Moran, J. M., Gabrieli, J. D. E., Adams, R. B. et al. (2010). Voting behavior is reflected in amygdala response across cultures. *Social Cognitive and Affective Neuroscience*, 5(2-3): 349–355. https://doi.org/10.1093/scan/nsp046.

Shahid, F., Zameer, A. and Muneeb, M. (2020). Predictions for COVID-19 with deep learning models of LSTM, GRU and Bi-LSTM. *Chaos, Solit. and Frac.*, 140: 110212.

Shen, G., Horikawa, T., Majima, K. and Kamitani, Y. (2019). Deep image reconstruction from human brain activity. *PLOS Computational Biology*, 15(1): e1006633. https://doi.org/10.1371/journal.pcbi.1006633.

Stallen, M., Borg, N. and Knutson, B. (2021). Brain activity foreshadows stock price dynamics. *The Journal of Neuroscience*, 41(14): 3266–3274. https://doi.org/10.1523/JNEUROSCI.1727-20.2021.

Zeroual, A., Harrou, F., Dairi, A. and Sun, Y. (2020). Deep learning methods for forecasting COVID-19 time-Series data: A Comparative study. *Chaos, Solit. and Frac.*, 140: 110121.

Zhao, S., Han, J., Jiang, X., Huang, H., Liu, H. et al. (2018). Decoding auditory saliency from brain activity patterns during free listening to naturalistic audio excerpts. *Neuroinformatics*, 16(3-4), 309–324. https://doi.org/10.1007/s12021-018-9358-0.

Chapter 8

Detection of COVID-19 in Lung CT-Scans using Reconstructed Image Features

Ankita Sharma and *Preety Singh**

ABSTRACT

COVID-19 is caused by coronavirus-2 (SARS-CoV-2) virus strain. The disease is easily communicable and fatal in nature. As it is contagious, its early and rapid detection is crucial to lower the death rate. In our research, we explore the use of reconstructed image features based on graph-based techniques for classifying the CT-scan images of Covid-19 suspected patients for detection of the infection. We also compare our proposed approach with the traditional approach of using images directly for classification with convolutional neural networks. Results show that our proposed approach using reconstructed features yields 99% accuracy along with high sensitivity.

Department of Computer Science & Engineering, The LNM Institute of Information Technology, Jaipur, India.
  Email: ankita.sharma.y19pg
* Corresponding author: preety@lnmiit.ac.in

1 Introduction

Coronavirus disease (Covid-19) is a communicable disease caused by the SARS-CoV-2 virus. It is a respiratory tract illness which can progress to acute respiratory failure and cause fatality if not diagnosed and treated in a timely manner (Li et al., 2020; Chen et al., 2020). Being fast progressing and contagious in nature, it requires swift identification and isolation to break the chain of infection, reduce its spread in the community and lower the mortality rate. Delay in the early segregation of an infected person can escalate the number of people getting infected.

The alarming proportion of casualties due to this global pandemic has resulted in many researchers doing extensive research to incorporate machine learning in the rapid diagnosis of the disease from radiological images (Ozturk et al., 2020; Naude, 2020; Gjesteby et al., 2017). Radiography examinations such as chest X-rays or computed tomography (CT) are done for examining the presence of the disease and its extent. The course of treatment can be better determined if a view of interior tissues is available, making the radiological images vital for accurate diagnosis.

However, in the event of the pandemic, when a large population has been affected, manual identification of the infection in CT-scans or X-ray images becomes humongous and tedious due to the substantial volume of patients. Manual handling of the substantial inflow of patient reports is a daunting task, labour-intensive, and prone to human glitches. Thus, the necessity of an artificially intelligent system for detection of the disease in lung images and to identify such a person is imperative due to the following reasons:

1. To rapidly identify the presence of Covid-19 in a suspected person.
2. To provide early treatment to the patient.
3. To arrange for immediate isolation of the patient for prevention of spread in the community.

The requirement of an automated model is to have an accuracy as high as possible while reducing false negatives and false positives. False negatives may result in infected people going undiagnosed, risking proliferation of the virus in society. False positives may cause unnecessary alarm and stress to people due to erroneous results. Many trained models have been proposed for detection of Covid-19 from CT-scan images with encouraging results. In our research, we propose an automated system to detect Covid-19 using reconstructed image features. These features are classified using artificial neural networks and compared with deep learning techniques like convolutional neural networks. The chapter is organized as follows: Section 2 outlines few related works in this domain. Section 3 presents our proposed methodology. Section 4 discusses the experiments and results. Section 5 concludes the chapter.

2 Literature Survey

Existing research in this field has shown that deep learning techniques are effective in identifying the positive Covid-19 patients as compared to other machine learning algorithms. In (Ozkaya et al., 2020), the authors have obtained patch images of the infected and non-infected regions from CT images. Feature vectors are obtained and correlation values between these features are considered for the fusion process. The T-test method is used to eliminate features close to each other and the fusion features are classified using a Support Vector Machine (SVM). The accuracy values reported for the subsets of the dataset are 95.60% and 98.27% respectively.

In (Ozturk et al., 2020), the authors extract robust features from images using different feature extraction techniques. The stacked Auto Encoder (sAE) architecture and Principal Component Analysis (PCA) are used for reduction of features and a SVM is trained for classification purposes. Classification results of shrunken features with sAE gives an accuracy of 71.92% while for shrunken features with PCA it is 94.23% with a sensitivity of 91.88%. Loey et al. (Loey et al., 2020) proposed a Conditional Generative Adversarial Nets (CGAN) for COVID-19 detection in chest CT scan images with a limited dataset. Five types of deep convolutional architectures are employed to detect the Coronavirus-infected patients. Results show that ResNet50 performs well in detecting COVID-19. The testing accuracy of the model is 82.91%, and the sensitivity and specificity values are 77.66% and 87.62% respectively.

In (Saha et al., 2021), a GraphCovidNet is proposed, which is a Graph Isomorphic Network based model to detect COVID-19 from CT-scans and X-ray images of the infected patients. The model shows an accuracy of 99% for both datasets. In (Liu et al., 2020), region of interest (ROI) delineation implementation is based on ground-glass opacities to accurately classify COVID-19. It uses gray-level co-occurrence matrix features, its gradient and histogram features. The authors report an accuracy of 94.16%, sensitivity of 88.62% and specificity of 100.00%. In (Wang et al., 2021), the authors have collected 1065 CT images of confirmed COVID-19 cases. Images which were earlier diagnosed with typical viral pneumonia are also included. Regions of interest (ROI) are marked based on the features of COVID-19 followed by manual delineation. A total accuracy of 89.5%, sensitivity of 0.87 and specificity of 0.88 is reported.

In (Yu et al., 2021a) a convolutional neural network is used for recognition of pneumonia. This model is based on graph-knowledge. The model is reported to achieve an accuracy of 0.9872, sensitivity of 1 and specificity of 0.9795. In (Ni et al., 2020), characterization of COVID-19 pneumonia in chest CT images is done. A deep learning algorithm is trained for lesion detection, segmentation and location. On a per-patient basis, the algorithm shows a superior sensitivity of 1.00 (95% confidence interval) and F1 score of 0.97 in detecting lesions. In (Yu et al., 2021b), ResGNet-C model using ResGNet framework is proposed. This is a graph convolutional network and applied to distinguish between normal and COVID-19 affected pneumonia images. Features are extracted from the ResGNet-C model. The Euclidean distance between the features is computed. This is followed by constructing graphs and encoding them to classify the CT images. The five cross-validation approach shows an averaged accuracy at 0.9662, sensitivity of 0.9733 and specificity at 0.9591.

3 Proposed Methodology

In our experiments, apart from utilizing the original image pixels, we have also worked with features reconstructed from pixel intensity values of the CT-scan images and classified them for detection of Covid-19. We have employed artificial neural network (ANN) and convolutional neural network (CNN) for classification purposes. We have applied the following approaches (refer Fig. 1):

- **M1:** Classification of Images using CNN
- **M2:** Classification of Images using Regulated CNN
- **M3:** Classification of Deep Image Reconstructed Features using ANN
- **M4:** Classification of Reconstructed Image Features using ANN

These approaches are discussed in the subsequent subsections.

M1: Classification of Images using CNN

In this approach, the pixel intensities of the images are directly used for classification of Covid-19 with a Convolutional Neural Network (CNN). To train the model on our limited training dataset, we apply transfer learning. The trained model is then tested on the test images.

M2: Classification of Images using Regulated CNN

In this approach, we first apply Principal Component Analysis (PCA) to the original images. It is observed that 2158 principal components correspond to 98% variance. Taking this as an indicator for the number of key features, the last fully connected layer of the CNN is configured to have 2158 neurons.

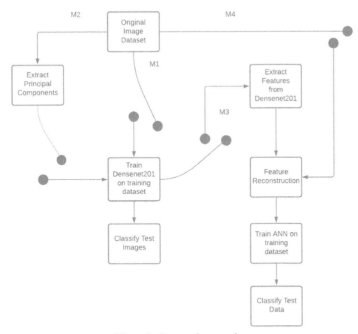

Figure 1: Proposed approaches.

This *Regulated CNN* is trained using the images in the train dataset and then used for classification of images in the test dataset.

M3: Classification of Deep Image Reconstructed Features using ANN

The CNN model is trained on the training set using transfer learning. Features are extracted from the top fully connected layer of the network. These *deep features*, extracted from the CNN, are reconstructed using the graph-based approach discussed in Section 3.1, and referred to as *Deep Image Reconstructed Features* and used for training an Artificial Neural Network (ANN). The trained ANN is then utilized for classification.

M4: Classification of Reconstructed Image Features using ANN

In this approach, the original pixel intensity values of the images are reconstructed as outlined in Section 3.1. These *Reconstructed Image Features* are used for training and testing an ANN.

3.1 Feature Reconstruction

The algorithm used for reconstruction of features has been proposed by Yu et al. (Yu et al. 2021a) and outlined below:

1. Divide features of all images into batches
2. For each batch of images, features of each image will form one node
3. Calculate the Euclidean distance between each node to the rest of the nodes. The Euclidean distance between two points p and q in an n-dimensional space is: $d(p,q) = \sqrt{\sum_{i=1}^{n}(q_i - p_i)^2}$
4. Find the k nearest neighbours for each node based on its similarity (distance) to the other points
5. Assign a value in the adjacency matrix at positions that represent the first k neighbours

Let N be the total number of images and S denote the batch size (number of images in each batch). Thus, the number of batches, n, will be:

$$n = \lceil S/N \rceil$$

The feature set of each image j is denoted by f_j and the ith batch of features is denoted by F_i. As the features extracted for each image is a column vector, each batch of images can be represented as:

$$F_i = [f_1, \cdots, f_j, \cdots, f_s]^T \quad (1 \leq j \leq S)$$

Thus, the set of all features of all images is:

$$F = \begin{bmatrix} F_1 \\ \cdots \\ F_i \\ \cdots \\ F_n \end{bmatrix} \quad (1 \leq i \leq n)$$

The next step is to compute the adjacency matrix, A_i, for each batch i. For a simple graph, this matrix contains values 1 or 0. A value 1 at position (m, n) indicates that nodes m and n are neighbouring nodes. A_i is computed as follows:

Step 1 Initialization: The initialization of variables used in the algorithm is done as follows:

- Adjacency matrix A_i of size S, initialized to zero:

  $A_i = Zeros(S, S)$

- Distance matrix *Distance* of size S, initialized to infinity:

  Distance $= Infinity(S, S)$

- Sorted distance matrix *SortedDistance* of size S, initialized to zero. This represents the nearest neighbours in ascending order:
  ***SortedDistance* = *Zeros*(S, S)**

- Index matrix *Index* of size S, initialized to zero. It shows the index of other nodes in order of the smallest distance from the current node:
  ***Index* = *Zeros*(S, S)**

Step 2 Distance calculation: For each image (node) p in a given batch, its distance is computed with every other image (node) q in the same batch using the following algorithm:

```
for p = 1:S
    for q = 1:S
        if p ≠ q
        Distance(p, q) = sum(f_p − f_q);
        end
    end
end
```

Assuming that each image has M features, the computation of the Euclidean Distance for feature sets f_p and f_q (corresponding to images p and q) in each batch F_i is as follows:

$$\text{Distance}(p,q) = \sqrt{\sum_{h=1}^{M}(f_p(h,1) - (f_q(h,1))^2}$$

Step 3 Output A_i: In this step, the computation of the values in the adjacency matrix is done, based on k nearest neighbours.

```
for p = 1:S
    SortedDistance(p,:), Index(p,:)] = sort(Distance(p,:))
    for c = 1:k
    A_i(p,Index(p, c)) = 1
    end
end
```

Each feature f_j in batch F_i is reconstructed as:

$$f_j = \tilde{A}_i^j F_i$$

where, $\tilde{A}_i^j$ is the *jth* row of $\tilde{A}_i$, the normalized adjacency matrix. The reconstructed feature batch F_i' can now be computed as:

$$F_i' = \tilde{A}_i F_i \qquad (1)$$

For normalizing the adjacency matrix A_i we calculate a degree matrix $D \in \mathbb{R}^{S \times S}$ as follows:

$$D(j,h) = \Sigma_{h=1}^{S} A_i(j,h)$$

This implies:

$$D(j,h) = \begin{cases} k, & if \ j = h \\ 0, & if \ j \neq h \end{cases}$$

We then normalize A_i as:

$$\tilde{A}_i = D^{\frac{1}{2}}(A_i + I)D^{\frac{1}{2}}$$

where, I is the identity matrix. Multiplying the normalized adjacency matrix $\tilde{A}_i$ with F_i gives us the reconstructed feature batch F'_i for classification using an Artificial Neural Network (ANN).

4 Experiments and Results

As presented in Section 3, classification of CT-scan images has been done using image features with and without feature reconstruction. The details of the dataset used in experiments and the classification results for different approaches are presented in the following subsections.

4.1 Dataset

For the CT-scan images, we have used the SARS-CoV-2 CT scan dataset (SARS-CoV-2CT-ScanDataset,2020). This contains 1252 Covid-19 positive and 1230 negative labelled images. All images are gray-level images. Some sample images depicting the progression of disease are shown in Fig. 2. As it is important to have a large dataset or deep learning techniques, the data was augmented to increase the existing dataset size. We used horizontal flipping which created mirror images of the existing images. This resulted in a dataset of 4800 images. The dataset was split into train, validation and test sets in the ratio 70:15:15.

4.2 Parameters of CNN and ANN Architectures

As the number of images in the available dataset is not large, we used a pre-trained CNN architecture and employed transfer learning using our training dataset. In all experiments where CNN was used for classification or for

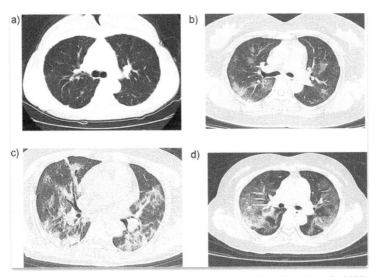

Figure 2: CT scan images showing progressive infection of Covid-19 (Kaggle, 2020).

feature extraction, we used the Densenet-201 architecture (Wang and Zhang, 2020) with the following hyperparameters:

- Number of layers - One fully connected (FC) layer and a classification layer were added as top layers to Densenet-201.
- Number of neurons in last FC layer - This depended on the number of features extracted from the last FC layer. In our experiments, we used 1024 and 2158 neurons.
- Activation function - Activation functions Relu was were used with the sigmoid function at the final classification layer.
- Loss function - Binary Cross-entropy loss function was used.
- Optimizer - RMSprop optimizer with a learning rate of 2e-5 was employed.
- Number of epochs - These varied from 10 (for 2158 features) to 20 (for 1024 features).

For the ANN used in our experiments, drop out layers were added after every fully connected (FC) layer to reduce overfitting. Below are the parameters used for the ANN in our experiments:

- FC layer with 128 neurons using activation function Relu.
- A drop out layer with 0.2 dropout rate.
- FC layer with 64 neurons using activation function Relu.
- A dropout layer with 0.2 dropout rate.

- FC layer with 32 neurons using activation function Relu.
- A dropout layer with 0.2 dropout rate.
- A final classification layer with sigmoid activation function.

For the purpose of classification of features into covid and non-covid, a probability value of 0.5 was set. A predicted value greater than 0.5 was labeled as 1 (Covid) and less than 0.5 as 0 (non-Covid) in the test dataset.

4.3 Evaluation Metrics

The metrics used for classification included accuracy, sensitivity, specificity, F1-score and Area under the ROC curve (AUC). Let *True Positives* be denoted by *TP*, *True Negatives* by *TN*, *False Positives* by *FP* and *False Negatives* by *FN*. Then, the evaluation metrics used are defined as:

$$Accuracy = (TP + TN)/(TP + TN + FP + FN)$$

$$Sensitivity = TP/(TP + FN)$$

$$Specificity = TN/(TN + FP)$$

$$Precision = TP/(TP + FP)$$

$$Recall = TP/(TP + FN)$$

$$F1 score = 2 * (Precision * Recall/Precision + Recall)$$

The results of various experiments conducted are discussed in the following subsections.

4.4 Results for Classification of Images using CNN

We trained Densenet-201 model on the training set using transfer learning. The top layer of the Densenet-201 was removed and replaced by one fully connected layer with 1024 neurons and a classification layer. The training and validation results of the CNN is shown in Fig. 3. Test images were classified using this trained model which resulted in an accuracy of 97.64%. The results are shown in the first row of Table 1.

Table 1: Train, validation and test accuracies for classification of images using CNN and Regulated CNN.

Classifier	Neurons in FC Layer	Train Accuracy (%)	Validation Accuracy (%)	Test Accuracy (%)
CNN	1024	99.75	98.01	97.64
Regulated CNN	2158	100	96.59	98.33

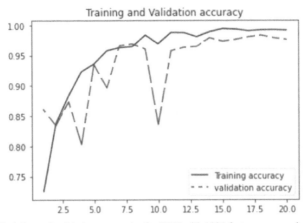

Figure 3: Training and validation accuracies for CNN with 1024 features vs. number of epochs.

4.5 Results for Classification of Images using Regulated CNN

Application of Principal Component Analysis showed 2158 principal components corresponding to 98% variance. The number of neurons in the final layer of the Densenet-201 architecture were fixed to 2158 and the model trained and tested with dataset images accordingly. The training and validation results of the CNN is are shown in Fig. 4. This approach resulted in an accuracy of 98.33%. The results using this regulated CNN are shown in the second row of Table 1.

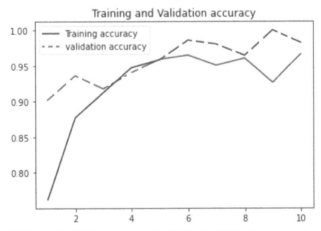

Figure 4: Training and validation accuracies for CNN with 2158 features vs. number of epochs.

4.6 Results for Classification of Deep Image Reconstructed Features using ANN

Transfer learning was applied to Densenet-201 by removing the top layers and introducing two new layers. A fully connected (FC) layer with 1024 neurons and activation function Relu was introduced. A classification layer with activation function sigmoid was then added. The 1024 features extracted from the FC layer were reconstructed as outlined in Section 3.1 for different values of n. The value of k was approximately set to the square-root of n. Results were also obtained for $k = 1$ and are shown in Table 2. As can be observed, highest accuracy of 96% using this approach was achieved for $n = 2400$ and $k = 1$.

Table 2: Evaluation metrics for classification of deep image reconstructed features using ANN.

n	k	Sensitivity (%)	Specificity (%)	F1-score (%)	Precision (%)	AUC (%)	Accuracy (%)
128	1	86	85.6	86	86	85.5	86
128	11	80	76.6	80	80	79.6	80
1200	1	93	93	93	93	93	93
1200	33	86.6	88	88	88	87.8	88
2400	1	96.6	96	96	96	96.1	96
2400	49	86.3	89	89	90	89.4	89

4.7 Results for Classification of Reconstructed Image Features using ANN

As the dataset image size is 250 × 250, the reconstruction of these pixel intensities values resulted in 62500 features which were used for training and testing using the ANN. The results for $n = 128$ with different values of k are shown in Table 3. For $k = 112$, an accuracy of 99% is achieved, along with a sensitivity of 99% and 100% specificity.

Table 3: Evaluation metrics for reconstructed image features using ANN.

k	Sensitivity (%)	Specificity (%)	F1-score (%)	Precision (%)	AUC (%)	Accuracy (%)
1	99	99.4	99	99	99.4	99
11	99	99.7	99	99	99.3	99
112	99	100.0	99	99	99.4	99

4.8 Discussion

In our experimental results, we achieve the highest accuracy of 99% using features reconstructed from pixel intensities and classified using an ANN (approach $M4$). If images are used directly, an accuracy of 97.6% is achieved with a convolutional neural network (approach $M1$), which improves to 98.3% if the CNN is regulated (approach $M2$). Reconstruction of deep image features and classification with ANN, results only in an accuracy of 96% (approach $M3$). The $M4$ approach of using reconstructed pixel intensities also results in high sensitivity which is required to decrease the number of False Negatives. Moreover, this approach requires less computation as a deep architecture is not used and a shallow neural network suffices. In comparison, the best accuracy achieved with a CNN is 98.3%. This could be due to non-availability of a large dataset for training the CNN.

In (Muller et al., 2020), a sensitivity of 94.1%, specificity of 95.5%, and Area under ROC curve (AUC) of 0.979 has been reported. (Liu et al., 2020) have shown an accuracy of 94.16%, sensitivity of 88.62% and specificity of 100.00%. In (Yu et al., 2021), an accuracy of 0.9872, sensitivity of 1 and specificity of 0.9795 is reported. Our approach shows considerable improvements in results by having a high accuracy of 99% along with a sensitivity of 99%. High sensitivity of our model contributes to a decrease in the number of False Negatives thereby ensuring that infected persons do not go undetected. This is a paramount consideration in the event of a pandemic where the illness is contagious.

5 Conclusions

In our proposed work, we have explored the use of reconstructed features and their classification with Convolutional Neural Networks and Artificial Neural Networks for the detection of Covid-19 in CT-scan images. We have compared the results with classifying the images directly with Convolutional Neural Network. Results show that reconstructing pixel intensity values and classifying them with an ANN gave a high accuracy of 99% and sensitivity of 99%. These results are enhanced as compared to classifying images with CNN. The high sensitivity also ensures that there are fewer False Negatives so that infected people are not missed out from being detected. This approach can be further explored for X-ray images as well.

References

Chen, N., Zhou, M., Dong, X., Qu, J., Gong, F. et al. (2020). Epidemiological and clinical characteristics of 99 cases of 2019 novel coronavirus pneumonia in Wuhan, China: A descriptive study. *Lancet*, 395(10223): 507–513.

Gjesteby, L., Yang, Q., Xi, Y., Zhou, Y., Zhang, J. et al. (2017). Deep learning methods to guide CT image reconstruction and reduce metal artifacts. pp. 752–758. *In*: Thomas G. Flohr, Joseph Y. Lo and Taly Gilat Schmidt (eds.). *Medical Imaging 2017: Physics of Medical Imaging*. International Society for Optics and Photonics, SPIE, 10132.

Li, K., Wu, J., Wu, F., Guo, D., Chen, L. et al. (2020). The clinical and chest CT features associated with severe and critical COVID-19 pneumonia. *Invest. Radiol.*, 55(6): 327–331. doi: 10.1097/RLI.0000000000000672.

Liu, C., Wang, X., Liu, C., Sun, Q., Peng, W. et al. (2020). Differentiating novel coronavirus pneumonia from general pneumonia based on machine learning. *Biomedical Engineering Online*, 19(1): 66. https://doi.org/10.1186/s12938-020-00809-9.

Loey, M., Manogaran, G. and Khalifa, N. E. M. (2020). A deep transfer learning model with classical data augmentation and CGAN to detect COVID-19 from chest CT radiography digital images. *Neural Comput. & Applic*. https://doi.org/10.1007/s00521-020-05437-x.

Naudé, W. (2020). Artificial intelligence against COVID-19: An early review. *Technical Report. IZA Inst. Labor Econ., Maastricht, The Netherlands*. https://www.iza.org/publications/dp/13110/artificial-intelligence-against-covid-19-an-early-review.

Ni, Q., Sun, Z. Y., Qi, L., Chen, W., Yang, Y. et al. (2020). A deep learning approach to characterize 2019 coronavirus disease (COVID-19) pneumonia in chest CT images. *European Radiology*, 30(12): 6517–6527.

Özkaya, U., Öztürk, Ş. and Barstugan, M. (2020). Coronavirus (COVID-19) Classification using deep features fusion and ranking technique. *In*: Hassanien, A. E., Dey, N. and Elghamrawy, S. (eds.). *Big Data Analytics and Artificial Intelligence Against COVID-19: Innovation Vision and Approach*. Studies in Big Data, vol 78. Springer, Cham. https://doi.org/10.1007/978-3-030-55258-9_17.

Öztürk, Ş, Özkaya, U. and Barstuğan, M. (2020). Classification of Coronavirus (COVID-19) from X-ray and CT images using shrunken features. *Int. J. Imaging Syst. Technol*. https://doi.org/10.1002/ima.22469.

Saha, P., Mukherjee, D., Singh, P. K., Ahmadian, A., Ferrara, M. et al. (2021). GraphCovidNet: A graph neural network based model for detecting COVID-19 from CT scans and X-rays of chest. *Scientific Reports*, 11: 8304.

SARS-COV-2 Ct-Scan Dataset, https://www.kaggle.com/plameneduardo/sarscov2-ctscan-dataset.

Wang, S. H. and Zhang, Y. D. (2020). DenseNet-201-Based deep neural network with composite learning factor and precomputation for multiple sclerosis classification. *ACM Transactions on Multimedia Computing, Communications, and Applications*, 16(60): 1–19.

Wang, S., Kang, B., Ma, J., Zeng, X., Xiao, M. et al. (2021). A deep learning algorithm using CT images to screen for Corona virus disease (COVID-19). *European Radiology*, 31(8): 6096–6104. https://doi.org/10.1007/s00330-021-07715-1.

Yu, X., Lu, S., Guo, L., Wang, S. H., Zhang, Y. D. et al. (2021b). ResGNet-C: A graph convolutional neural network for detection of COVID-19. *Neurocomputing*, 452: 592–605. https://doi.org/10.1016/j.neucom.2020.07.144.

Yu, X., Wang, S. H. and Zhang, Y. D. (2021a). CGNet: A graph-knowledge embedded convolutional neural network for detection of pneumonia. *Information Processing & Management*, 58(1): 102411. https://doi.org/10.1016/j.ipm.2020.102411.

Chapter 9

Dental Image Analysis: Where Deep Learning Meets Dentistry

Bernardo Silva,[1] *Laís Pinheiro,*[1] *Katia Andrade,*[2]
Patrícia Cury[2] *and Luciano Oliveira*[1,*]

ABSTRACT

Deep learning has profoundly advanced many fields. In computer vision, this progress became feasible due to a combination of immense computing power, large data availability, and convolutional neural networks (CNNs). CNNs were a game-changer for computer vision because they efficiently process grid-like data and adapt the final model's behavior according to the target. With these means, today's researchers are now proposing tools and methods that were im-

---

[1] Intelligent Research Vision lab, Federal University of Bahia, Bahia, Brazil.
[2] Program in Dentistry and Health, School of Dentistry, Federal University of Bahia, Salvador, Bahia, Brazil.
* Corresponding author: lrebouca@ufba.br

practicable and seemed unreal a decade ago. Many methods and tools were and are being proposed and developed in the medical imaging field. However, the deep learning applications in dentistry are still incipient, being an open-field for studies and researchers entering the field. This chapter offers a comprehensive discussion on the most common images used in dentistry and their deep learning literature applications, relating them to the corresponding computer vision task. We discuss promising avenues to the complex problem of automatic report generation. At the end of this chapter, we present an overview of the field's achievements, opportunities, and challenges.

1 Introduction

Much has been said in the media about the fascinating advances in artificial intelligence (AI) applied in several fields, especially when involving image and video footage analysis on people analytics or surveillance. In the last decade, AI has boomed because of a new study field created through machine learning techniques, so-called deep learning. In this new research field, the learning process takes place on artificial neural networks (ANNs) that, conversely to the way these networks did traditionally, explore the use of several hidden layers to extract features after being appropriately trained. In deep neural networks (DNNs), feature extractors are trainable and found across convolutional operations with kernel matrices composed of weights defined in the training process. Hence DNNs are mainly based on convolutional neural networks (CNNs), although other techniques (Vaswani et al., 2017), which do not explore convolutions, have gained more room, showing a superior performance in many cases.

CNNs benefit from general processing units (GPUs) and big data. It happens as CNNs own several parameters to be tuned during the training process, consequently dependent on powerful computational resources to run optimization procedures and considerable volumes and varieties of training data. Particularly in health, the application of CNNs reaches a consensus both because of the availability of large amounts of training data and their high performance ability to execute proposed tasks. In dentistry, for example, even though it is not yet possible to come across such robust applications in clinical routines, it is possible to find available several pre-clinical works in the literature that illuminate the path of AI towards clinical practices (Silva et al., 2018; Jader et al., 2018; Silva et al., 2020).

In this chapter, we thoroughly discuss how research on CNNs has been conducted in dentistry. In Section 2, we address the habitual images used in the dentist's daily routine to support our discussion. Section 3 presents general essential aspects of computer vision (CV) tasks that are supported by CNN architectures and used on the types of radiographs discussed in the previous section; some essential state-of-the-art works are also contextually considered in this section. Section 4 approaches a relevant future perspective of generating automatic dental reports from panoramic radiographs. Finally, Sections 5 and 6 conclude this chapter by delineating current pre-clinical achievements, still-opened challenges, and concluding remarks.

2 Images in Dentistry

Oral health is a fundamental indicator of general health and quality of life. Many diseases and conditions affect people's health, including dental caries, periodontal diseases, other infections, traumas in the maxillofacial region, neoplasias, and congenital disabilities, such as cleft lips and palates. The Global Burden of Disease Study 2017 reckoned that oral diseases impact around 3.5 billion people globally (WHO. https://www.who.int/health-topics/oral-healthtab=tab1).

Images are fundamental components for diagnosing oral diseases and conditions, planning treatments, and monitoring the patient's maxillofacial infirmities, being also essential for forensic dentistry. Technological advances have increasingly raised the importance of image use, improving the diagnosis while allowing the analysis of characteristics or health conditions. Technology also improved the communication between health professionals and the patients, helping to educate the population and prevent diseases.

In dentistry, different types of images are acquired every day world wide. Clinical and histological photographs and conventional dental radiographs (X-rays) account for the bigger fraction of all images acquired. Three-dimensional images have also been used, although they are beyond the most general dental demands. Among these, the most common are computed tomography and cone-beam imaging. Over 1.4 billion dental X-rays obtained in 2018 in the United States of America include intraoral and extraoral radiographs and cone-beam computed tomographies, more than 90% being intraoral radiographs (IDATARESEARCH. https://idataresearch.com; Brown, 2001; Jayachandran, 2017). Intraoral and extraoral scanning, magnetic resonance imaging, nuclear medicine, and ultrasonography are types of equipment generally used for more specialized applications, presenting higher costs.

A radiograph is a two-dimensional image obtained from a three-dimensional structure. In this technique, the complete volume of the target tissue is projected onto a two-dimensional image. The digital radiographic technique uses receptors, such as wired, wireless, or phosphor plate sensors instead of plastic films. In digital imaging, the photons from the X-ray tube head reach the sensor, and the analog image is converted to digital and transferred to a program. Digital images show better quality than conventional ones. They can be interactively manipulated after the acquisition and transferred to computational analysis or other health professionals without altering the original quality. Digital receptors require 90% less radiation than film, reducing exposure to ionizing radiation (White and Pharoah, 2008; White and Pharoah, 2009; Sabarudin and Tiau, 2013). Thus, digital imaging has revolutionized the radiology field, making images suitable for computational applications such as CNNs.

Radiographs are indicated when they offer auxiliary diagnostic information and influence the treatment plan. Clinical signs or symptoms in the patient's history will be indicated when a radiologic examination is required. A clinician must clearly understand the normal maxillofacial anatomy to interpret a radiograph and then mentally reconstruct a three-dimensional image from two-dimensional views. This procedure requires extensive training, which is frequently time-consuming (White and Pharoah, 2008; White and Pharoah, 2009; Sabarudin and Tiau, 2013). The two main types of radiographs are the intraoral (the radiographic film or the sensor is set inside the mouth) and extraoral (the radiographic film is placed outside the mouth). For both radiographs, the X-ray source is outside the mouth.

Intraoral X-rays are the type used most and are considered the foundation of imaging for the dentist. They are categorized into periapical, bitewing, and occlusal. Periapical radiographs show the entire tooth and the surrounding bone (see Fig. 1(a)). They allow the assessment of root canal morphology, alveolar bone level in the interproximal region, periapical pathology, and tooth fractures. Bitewing radiographs show upper and lower crowns of teeth and a small portion of bone crests (see Fig. 1(b)). These types of radiographs are used for detecting interproximal and secondary caries and bone loss in the early stages of periodontitis. Occlusal radiographs show a horizontal projection of teeth and a part of the bone (Fig. 1(c)). They are used to find the location of supernumerary and impacted teeth, foreign bodies in the maxillofacial region, and salivary glands stones.

Extraoral radiographs are used to examine more extensive areas of the maxillofacial region, monitor growth and development of the skeleton and or-

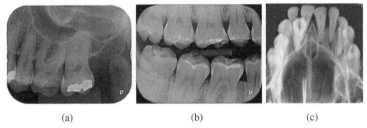

(a)   (b)   (c)

Figure 1: Intraoral X-rays: (a) Periapical radiograph. Maxillary molar region showing all of the molars and the surrounding bone. (b) Bitewing radiograph. The crowns and alveolar crest of maxillary and mandibular molars. (c) Occlusal radiograph. Anterior maxilla, showing the anterior floor of the nasal fossa and canines and incisors.

thodontic therapy, locate impacted teeth and large pathological lesions, and evaluate the temporomandibular joint, maxillary and mandibular fractures. The most common extraoral radiography is the panoramic one, which provides an overview of the dental arches and the surrounding maxillofacial structures, such as the maxillary sinuses and the temporomandibular joint (Fig. 2(a)). Panoramic radiographs have also been evaluated as a cost-effective tool to determine bone mineral density. They are easy and fast to take and associated with low patient radiation doses. Panoramic radiographs can be performed in patients whose mouth opening is limited or with disabilities. Patients readily understand panoramic images, and hence this type of radiograph is also a valuable visual aid in education and case presentation. Shortcomings of the panoramic radiographs are the geometric distortions and the lack of fine details available on intraoral radiographs (White and Pharoah, 2009).

Lateral teleradiography of the head is also one of the most frequently used images in dentistry. Skeletal, dental, and soft tissue anatomic landmarks are used to measure and categorize patient craniofacial morphology for orthodontic and orthognathic therapy. Dentists compare these measurements with a standard or compare the pre and post-treatment status to monitor treatment and patient growth (see Fig. 2(b)).

In 3D imaging, a series of anatomical data is obtained using specific equipment, processed by computer software, and visualized on a 2D monitor displaying the illusion of deepness. Currently, the most used methods are Computerized Tomography (CT) and Cone Beam Computerized Tomography (CBCT) (White and Pharoah, 2008; White and Pharoah, 2009; Shah et al., 2014).

CT was the first technique that allowed visualization of the maxillofacial region's hard and soft tissues due to image processing enhancement and the obtention of many non-superimposed cross-sectional images. It uses a narrow fan-shaped X-ray beam around a target anatomical structure to reveal its

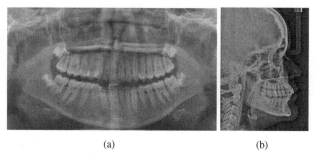

(a)   (b)

Figure 2: Extraoral X-rays: (a) Panoramic radiograph. The maxillary and mandibular dental arches, and their supporting structures. (b) Lateral teleradiography of the head. Skeletal, dental, and soft tissue anatomic structures.

interior content, allowing three dimensional visualization by the professionals. Current CT scanners, called multi-slice CT, have a linear array of up to 64 detectors that simultaneously obtain tomographic data at different section locations. However, these scanners are expensive and have limited application in maxillofacial diagnoses, like detecting injuries and tumors. CT software can highlight lesions from normal anatomic tissues using different colors and show slices of the target tissue with different slice thicknesses (1–2 mm) chosen by the professional. However, CT imaging causes a high radiation exposure to the patients and has poor resolution (White and Pharoah, 2008; White and Pharoah, 2009; Shah et al., 2014). Figure 3 shows the visualization of a CBCT image together with its 3D reconstruction.

The CBCT imaging technique utilizes a cone-shaped radiation source and a two-dimensional x-ray detector fixed on a rotating gantry, allowing the acquirement of multiple sequential images during a complete scan around the target area. This imaging technique produces a series of 2D images, which can be observed individually or reconstructed in 3D images using an algorithm (Fig. 3). The main advantages of CBCT over CT are the lower radiation doses, lower cost and a higher resolution. CBCT has numerous applications in all dentistry fields, namely endodontics, periodontics, implantology, orthodontics, and oral and maxillofacial surgeries, helping diagnose cysts, tumors, infections, bone loss, developmental anomalies, and temporomandibular joints and traumatic lesions (White and Pharoah, 2009; Shah et al., 2014).

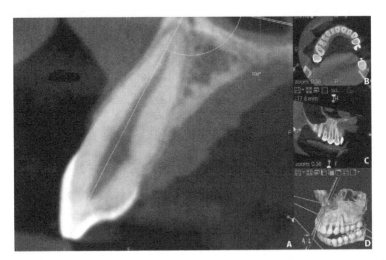

Figure 3: CBCT image. Sections of the three reference planes: A-Sagittal, B-Axial, C-Frontal, D-3D reconstruction.

3 Analysis of Dental X-rays through CNNs

The computer vision field has drastically advanced and changed in the years gone by primarily due to CNNs, which operate on "grid-like" data, such as images and CT scans (Goodfellow et al., 2016). CNNs are not a new technique: They were introduced by Yann Lecun (LeCun et al., 1998) and were successfully applied in digit classification. It was only in 2012, when the AlexNet architecture (Krizhevsky et al., 2012) won the ImageNet Large Scale Visual Recognition Challenge (ILSVRC) (Deng et al., 2009), that became notorious for its potential. AlexNet stacks eight neural network layers (five convolutional layers and three fully connected layers) and surpass its contenders by a significant margin. At that time, eight layers were considered to give the architecture a deep structure, in contrast with the shallow two, three layer counterparts. Since then, CNNs have prevailed in computer vision.

Convolution layers exhibit unique properties that make them suitable for image processing, such as shift-invariance, sparsity of connections, and weight sharing. In the end, a convolutional layer yields better results using just a fraction of the weights that fully connected layer solutions would use. The "convolutional" term refers to the namesake discrete mathematical operation, although the latter is slightly different from the one used in computer vision. The convolution operation is linear, being ineffective in stacking many convolutional layers to enhance network performance without adding non-linearities. Indeed, a non-linear activation function should follow each convolutional operation to expand the network capabilities, being the central idea of deep learning.

A neural network training procedure should follow one learning class: Supervised or unsupervised. For deep learning applications, a large amount of data is essential for both classes, but the supervised learning techniques require labeled data. Unsupervised learning does not use labels because it pursues data clustering through common patterns; supervised learning uses the labels as correct answers (ground-truth), adjusting the model parameters at each training step towards the error reduction direction.

The labeling procedure is usually manual and often time-consuming, sometimes exceeding 90% of the implementation time. Even with additional labeling costs, most researches and applications employ supervised learning methods due to their current superior performance in most cases. The labels are in accordance with the computer tasks, which include image classification, object detection, semantic segmentation, instance segmentation, and key-point detection. Figure 4 illustrates the previously discussed dental X-ray images, some corresponding tasks commonly performed by clinical dentists, and an example

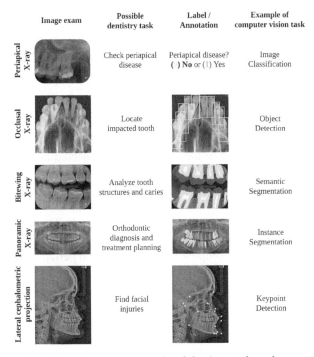

Figure 4: Some dental imaging exams, associated dentistry task, and an example of a CV task over the corresponding image.

of a CV task that might be accomplished over each type of X-ray[1]. Next, we discuss each computer vision task and review some works that explore dental applications using deep learning techniques for each task.

3.1 Image Classification

In image classification, a model must assign a category to an entire image. The labeling procedure is relatively inexpensive, and an annotator must label an image according to its category. For instance, an image may be numerically labeled as "1", representing the presence of a tooth fracture, or "0" otherwise. Figure 4 displays an example of a classification label for a periapical X-ray regarding the presence or absence of periapical disease.

There are several CNN architectures for image classification. The first successful ones were LeNet-5 (LeCun et al., 1998) and AlexNet (Krizhevsky et al., 2012), the former presenting only five layers. The VGG architectures (Simonyan and Zisserman, 2014) are also noteworthy and known for their modular and straightforward implementation. The Inception architectures (Szegedy

---

[1] Although all CV tasks can be performed on each type of X-ray image, in the figure, we illustrate just one task per image.

et al., 2015; Szegedy et al., 2016) are more complex than those for VGG but have fewer parameters. More recent networks such as ResNet (Kaiming et al., 2016) and EfficientNet architectures (Mingxing and Quoc, 2019) are reasonable choices for various problems.

There is no specific rule that guarantees satisfactory classification performance, but the architectures typically use numerous modular blocks consisting of small kernel convolutional layers followed by a pooling layer. These blocks are stacked with the goal of reducing the feature map width and height while increasing depth. At the top of the CNNs, fully connected layers classify the images. Figure 5 illustrates a Lenet-5 like architecture, which follows the described pattern.

In dentistry, we find several studies that proposed methods whose primary goal is to classify an image or part of it. Karatas et al. (Ozcan et al., 2021) use a Resnet-34 to differentiate amalgam, composite resin, and metal-ceramic restorations. Their data set comprises 550 bitewing and periapical radiographs, from which 110 images were used for testing. Poedjiastoeti et al. (Wiwiek and Siriwan, 2018) employed a VGG-16 to classify jaw tumors into ameloblastomas or keratocystic odontogenic tumors. The network was trained and validated on 400 panoramic radiographs, and its performance assessing 100 images was comparable with that of an oral maxillofacial specialist. Miki et al. (Yuma et al., 2017) numbered CT tooth crops using AlexNet. They performed their experiments on 52 CT volumes, from which 10 images were used for assessment, and reached 88% accuracy. Lee et al. (Ki-Sun et al., 2020) investigated diagnosing osteoporosis using crops of panoramic radiographs. The neural network architecture was a VGG-16, and the dataset consisted of 680 panoramic radiographs. The network accuracy was 84% on 136 test images.

Instead of manually cropping the input image, some of these methods could detect the regions of interest and then perform classification. This task is called object detection.

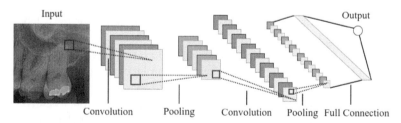

Figure 5: Convolutional layers followed by polling layers are a typical pattern in classification architectures. The feature map width and height reduce as we go forward in the network, while the depth increases. In the end, fully connected layers finally provide the image classes.

3.2 Object Detection

In object detection, a model delimits objects of interest using bounding boxes and possibly classifying them. The annotation procedure in object detection is more time-consuming than in classification as the bounding boxes are defined by two points, and a single image may contain multiple objects. Figure 4 illustrates an example of bounding box annotations in an occlusal image, detecting the teeth, and classifying them as impacted or not.

Object detection is a challenging task because the objects may vary in number, class, and size. Consequently, the CNN object detection architectures are more complex than the classification ones. There are fundamentally two detector families: One-stage or two-stage. The one-stage detectors derive from the YOLO architecture (Joseph et al., 2016; Joseph and Ali, 2017; Joseph and Ali, 2018) and can perform real-time detection. The two-stage detectors follow the pipeline introduced by Faster R-CNN (Ren et al., 2015). For medical image analysis, Faster R-CNN is usually preferable to YOLO because two-stage detectors allow larger inputs, are more precise, and the addition of a segmentation branch is straightforward.

Figure 6 illustrates a two-stage object detection architecture. The first stage consists of a backbone (a classification architecture such as the one previously mentioned) to extract the image features. A region proposal network (RPN) processes these features, outputting regions of interest (RoIs), each one possibly containing an object. A pooling technique, such as the RoIAlign, resizes the RoI features to a fixed size. In the second stage, the high-score RoIs bounding boxes are refined and classified by means of fully connected layers.

Object detection is a common task in digital dentistry. Chen et al. (Hu et al., 2019) trained a Faster R-CNN with 1,000 images to detect and classify teeth in periapical radiographs. They accessed the model on 250 images, reaching a mIoU above 90%. Kats et al. (Lazar et al., 2019) investigated atherosclerotic carotid plaque detection using crops of panoramic radiographs and a Faster R-

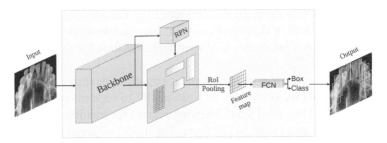

Figure 6: Representation of an object detection Faster R-CNN-like architecture. The first stage defines possible regions with objects of interest. The second stage confirms the presence of these objects, classifies them and refines the bounding boxes.

CNN. The data set consisted of 42 training images and 13 testing images, and the model reached 83% accuracy. Thanathornwong and Suebnukarn (Bhornsawan and Siriwan, 2020) employed a Faster R-CNN to detect periodontally compromised teeth on panoramic radiographs. The network training procedure used 80 images, and the model F_1 score on 20 test radiographs was 81% for a 0.5 IoU threshold. Kwon et al. (Odeuk et al., 2020) used a YOLOv3 to detect and classify odontogenic cysts and tumors on panoramic radiographs. The data set comprised 1,282 images, of which 20% were for assessment. The proposed model reached 95.6% accuracy.

When one is not willing to localize objects through bounding boxes but rather identifying their pixels, the task is segmentation. However, it is unreasonable to consider some structures as objects since they can not be separated into instances. In these cases, it is more appropriate to segment them according to their category, performing semantic segmentation.

3.3 Semantic Segmentation

Semantic segmentation can be seen as a more challenging classification task since it requires the classification of each pixel. The labeling procedure is costly and usually done through specialized annotation tools. The annotator must click around the borders as accurately as possible to segment the regions of interest. Consequently, the semantic segmentation data sets are scarce and small, making them very valuable. We illustrate in Fig. 4 a semantic segmentation annotation for the bitewing image, where each color represents a structure.

As in the other tasks, there is a large variety of semantic segmentation architectures. The most prominent ones are the fully convolutional network (FCN) (Long et al., 2015), DeepLab (Chen et al., 2017; Chen et al., 2017), and U-Net (Ronneberger et al., 2015), the latter being the most common choice for medical image analysis. The U-Net architecture consists of two paths in an encoder-decoder fashion: The contracting path that employs convolutions and the symmetric expansive path that employs up-convolutions. Moreover, the network contains skip-connections, concatenating features of the contracting path with the features of the expansive path. The encoder-decoder format and the skip-connections result in a U-shaped representation for the architecture, as depicted in Fig. 7.

Segmentation architectures are applied largely in dentistry. Cantu et al., 2020 used a U-Net backboned by an Efficient-B5 to segment caries in bitewing radiographs. They used 3,545 bitewing images for training and validation and assessed the model with 141 of them. The model achieved 81% accuracy, which showed it was better than the dentist invited to participate in the experiments. Setzer et al., 2020 segmented periapical lesions and other structures on CBCT images using a U-Net. The data set comprised of 20 CBCT volumes,

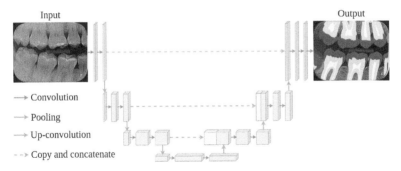

Figure 7: A U-Net-like architecture. The architecture consists of two almost symmetrical paths in an encode-decoder fashion. These paths are linked by skip connections.

and the model reached 91% of accuracy on the 4 test volumes. Jaskari et al. (Jaskari et al., 2020) segmented mandibular canals on CBCT scans using a 3D U-Net-like architecture. The data set consisted of 637 coarsely annotated volumes and 15 accurately annotated ones. They reached 57% of the dice similarity coefficient, making the authors conclude that the proposed approach could reduce the manual labor time for segmenting mandibular canals.

Semantic segmentation has proven to be applicable in dentistry when the structures are not associated with a single object. If not the case, one may do instance segmentation.

3.4 Instance Segmentation

In the instance segmentation task, a model pursues to segment and classify each object in a scene. It is an extremely challenging task since it combines object detection and semantic segmentation tasks. Instance segmentation should come to mind when segmentation should occur on unambiguous instances, such as persons and cars, while semantic segmentation should be preferable on structures without an explicit instance, such as streets and sky. As in semantic segmentation, the labeling procedure is quite costly because the annotator must classify the objects and segment them as accurately as possible around their borders. Figure 4 displays an instance segmentation annotation on a panoramic radiograph, in which each tooth was individually segmented and classified.

Traditional instance segmentation architectures rely mostly on object detection architectures, Mask R-CNN being the most known architecture (Kaiming et al., 2017). Mask R-CNN extends Faster R-CNN by adding a FCN segmentation head, as shown in Fig. 8. The Mask R-CNN also introduces the RoiAlign technique, a quantization-free layer that accurately preserves spatial information. Mask R-CNN performance can be improved by proposing novelties to the backbone (Zhang et al., 2020), the segmentation head (Kirillov et

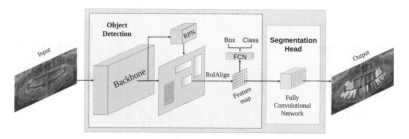

Figure 8: The Mask R-CNN architecture extends the Faster R-CNN by adding a segmentation branch, also known as the segmentation head.

al., 2020), the feature pooling technique (Liu et al., 2018), or even by changing the information flow at the second stage (Chen et al., 2019).

Chang et al. (Chang et al., 2020) use a Mask R-CNN to segment periodontal bone level, the cementoenamel junction, and the teeth as a preprocessing step to diagnose periodontal bone loss and stage periodontitis. They performed their experiments on a total of 330 periodontal bone levels, 115 cementoenamel junctions, and 73 tooth images. The intra-class correlation coefficient attested that the method achieved excellent performance. Boiko et al. (Boiko et al., 2019) investigated instance segmentation on 116 dental hyperspectral images. They employed a Mask R-CNN backboned by a ResNet-101 and obtained promising results for a few classes.

Other dentistry instance segmentation studies are primarily on panoramic teeth X rays, which we discuss later. The amount of teeth segmentation studies comes from the fact that each mouth has numerous instances, and detecting and segmenting them is essential for many tasks.

3.5 Keypoint Detection

Keypoint detection consists of localizing a set of predefined points in an image. In its simplest version, as it usually occurs in dentistry, the number of points is fixed. A more complex kind would consist of localizing a set of keypoints per object. For keypoint detection, the annotator must indicate where the related points must be located. Figure 4 illustrates keypoint labels depicted on a lateral cephalometric projection, which is a typical task performed by radiologists.

When the number of points is fixed, the keypoint detection architectures do not differ significantly from the classification architectures. The core difference is that they work as a multi-output regression-regression problem, i.e., the network outputs are continuous values instead of categorical. In an image, there will be two outputs per keypoint, as required to determine its location. Therefore, it is possible to tackle a keypoint detection problem by just modifying the last layer of a classification architecture, as illustrated in Fig. 9.

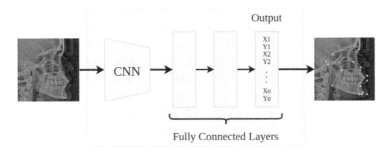

Figure 9: Architecture for a fixed-number keypoint detection. In this case, the architecture does not significantly differ from a classification one.

Keypoint detection is an exact example of a task performed by radiologists in cephalometric X-rays. Lee et al. (Lee et al., 2017) investigate two CNN solutions to detect 19 keypoints (38 coordinates) on cephalometric X-rays. They performed the training of the models on 150 images and assessed them on others, 150, showing promising performance. Song et al., 2020 propose a two-stage solution to detect the cephalometric keypoints. The first stage creates patches from coarse locations from the landmarks. The second stage precisely locates the landmarks through a ResNet-50, surpassing most methods that employed the same data. Chung et al. (Chung et al., 2021) approach the tooth detection problem as a keypoint detection: A point is localized for each permanent tooth, despite its presence. This process automatically labels the teeth.

4 Automatic Report Generation on Panoramic X-rays

Among dental image exams, panoramic X-rays stand out for their versatility. These images vary in composition and display many anatomical structures with considerable overlapping, as it shows the entire dental arch in a frontal view, making them also challenging. Consequently, most computer vision studies had neglected panoramic X-rays in the past. Silva et al., 2018 did a comprehensive systematic review on segmentation methods applied to dental X-rays. They found out that works on unsupervised handcrafted feature extractors before 2018 usually focused on intraoral images due to their simple nature. Concurrently, Silva et al. showed that a Mask R-CNN architecture could by far outperform all the traditional segmentation methods over panoramic images. Schwendicke et al., 2019 reviewed CNN works proposed until August 2019 for dental image diagnostics and found that panoramic X-rays were the most common study objects. These remarks revealed that the advent of CNNs allowed researchers to tackle and propose tools for more relevant and demanding problems.

The CNN applications for panoramic X-rays include age and sex estimation (Avuçlu and Basçiftç, 2020), caries detection (Vinayahalingam et al., 2021), osteoporosis diagnosis (Lee et al., 2021), and forensic identification (Gurses and Oktay, 2020). However, the ultimate goal for an automatic panoramic X-ray analysis is to write a clinically accurate report. For this purpose, a machine learning model should receive an input image and generate a textual output. This task is entirely different from the ones previously presented, and it is commonly referred to as "image captioning".

Image captioning is noticeably an arduous task, as it mixes computer vision and natural language processing (NLP). It allows supervised learning through the use of texts as the ground truth. The labeling procedure may consist of collecting textual descriptions for each image, which is expensive. Opportunely, there is no need to annotate medical images to solely train machine learning models, as the radiologist's textual reports are the ground truth wanted.

A question that must arise on the end-to-end supervised learning approach to report generation is whether this strategy will yield a satisfactory result or not. We are skeptical of this approach since panoramic radiograph analyses require the execution of essential tasks that radiologists perform unawares when writing a report. One of such task is tooth recognition.

4.1 Tooth Detection, Numbering, and Segmentation

It is expected that an individual with a healthy mouth during a lifetime forms 52 teeth: 32 permanents and 20 deciduous, all of them named according to their function and location, making their names very long. For this reason, numerous notations were proposed to expedite communication and form filling. The most common notation is the FDI World Dental Federal, in which a two-digit number represents each tooth. The first digit specifies the tooth's quadrant and dentition type (permanent or deciduous), while the second digit specifies the tooth type according to the dentition. Figure 10 illustrates the FDI notation along with an added color code for visualization purposes. Note that the right side corresponds with the left axis because, as in panoramic radiographs, we visualize the dentition from the front.

Tooth segmentation and numbering is a necessary pre-processing step for many automated radiograph analysis methods. Specifically for panoramic radiographs, teeth recognition is essential for clinical dentists for the following reasons: (i) a written report must document the patient's missing teeth, (ii) the radiologists use the teeth as landmarks when screening a panoramic X-ray, and (iii) many report findings use tooth locations and numbering as reference points. Furthermore, tooth numbering and segmentation are fundamental to forensic identification.

In the literature, we find works that perform tooth semantic segmentation prior to the spreading of DCNN. Most of these works relied on designed filters,

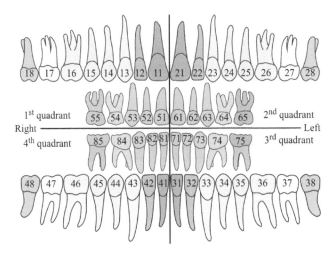

Figure 10: Representation of the two-digit FDI Notation System. The first digit specifies the tooth quadrant and the dentition type while the second digit refers to the tooth type.

i.e., a person would manually define the kernel size and weights accordingly to the considered task. These filters are called "handcrafted feature extractors" and usually represent some image characteristics, such as the object borders. The borders are certainly a valuable resource for detecting and segmenting objects. Still, they are barely the primary goal of an application, and, specifically for medical X-rays, the low contrast can be prohibitive for the use of such filters. Indeed, Fig. 11 illustrates the border detection on a dental panoramic image using the Sobel operator, disclosing that it failed to detect relevant borders. Silva et al., 2018 investigated several handcrafted feature extractors for segmenting teeth on dental panoramic X-rays, including the Sobel operator. They concluded that none successfully segmented the teeth, exposing a major limitation of handcrafted solutions. The main advantage of CNN models over handcrafted feature extractors is that they can fit adequate filter weights from the data accordingly to the task instead of employing handpicked ones.

Silva et al., 2018 were pioneers in applying deep learning to segment teeth in panoramic radiographs, although in a semantic fashion. They employed a Mask R-CNN network, which outperformed all investigated handcrafted feature extractor solutions. Their experiments used a high variability data set comprising 1,500 panoramic radiographs, which they made publicly available under the name UFBA-UESC Dental Images data set[2]. Jader et al., 2018 extended their previous work, also employing Mask R-CNN, by segmenting tooth instances on a subset of 276 images of the UFBA-UESC Dental Images data

---

[2]Instructions on how to request the data set are available at https://github.com/IvisionLab/deep-dental-image.

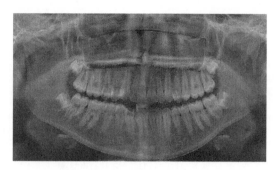

Figure 11: Border detection on a dental panoramic X-ray through a handcrafted feature extractor.

set[2]. The instance segmentation was only possible by manually modifying the semantic binary masks, disregarding tooth overlapping. Tuzoff et al., 2019 detected and numbered teeth on panoramic X-rays, without segmenting them, using a two network solution. The first network, a Faster R-CNN, detects the teeth. Crops are created according to the detection, and they feed the second network, a VGG-16, which classifies the teeth. Their experiments used 1,574 panoramic radiographs, and their solution was close to an expert level. Silva et al., 2020 investigated several end-to-end deep neural network architectures (Mask R-CNN, PANet, HTC, ResNeSt) for segmenting and numbering teeth. They used 543 radiographs with only permanent teeth of the UFBA-UESC Dental Images data set, modifying the binary masks from Silva et al., 2018 and numbering each tooth[3]. The PANet had the best performance, and the results attest to the feasibility of segmenting and numbering teeth on panoramic X-rays by end-to-end solutions. Finally, Pinheiro et al., 2021 annotated from scratch 450 panoramic radiographs of the UFBA-UESC Dental Images data set, considering overlapping and deciduous teeth, which were neglected elements in previous works[3]. They also refined the predictions by using the PointRend Module as the segmentation head of Mask R-CNN architecture. Figure 12 illustrates the advancement in tooth segmentation in the aforementioned works.

4.2 Architectures for Image Captioning

Early image caption architectures would employ a CNN, i.e., VCG, Inception as an image feature extractor and a recurrent neural network (RNN) to output text (sequential data), as depicted in Fig. 13 (Vinyals et al., 2015; Xu et al., 2015). The common choice for a RNN was the long short-term memory (LSTM) (Hochreiter and Schmidhuber, 1997), as its memory cells lead to

---

[3]Instructions on how to request the data set are available at https://github.com/IvisionLab/dns-panoramic-images.

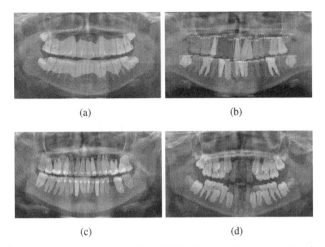

Figure 12: Advances in tooth segmentation. (a) Tooth semantic segmentation by Silva et al., 2018 (b) Tooth instance segmentation without numbering by Jader et al., 2018. (c) Permanent tooth segmentation and numbering by Silva et al., 2020. (d) Permanent and deciduous segmentation and numbering considering overlapping by Pinheiro et al., 2021.

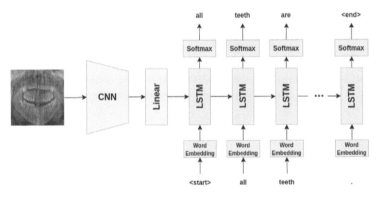

Figure 13: An early image captioning architecture. The CNN extracts image features, which go to a LSTM network to produce text output.

better performance on long sequences. These architectures could benefit from transfer learning from image data sets, such as the ImageNet (Deng et al., 2019), and have predictions refined by attention mechanisms (Xu, 2015).

The deep learning development significantly changed the state-of-the-art architectures for image captioning, much as a consequence of the Transformer architectures (Vaswani et al., 2017). These architectures dispense totally convolutional layers and recurrence, relying only on attention mechanisms and positional encodings. The first application of the transformer architecture was on translation (a sequence-to-sequence task), but it quickly became notorious with the transformer's potential for other applications. Nowadays, transform-

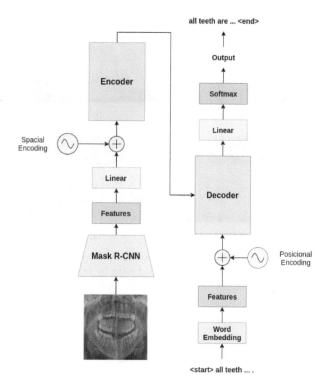

Figure 14: Image features from the encoder should feed the decoder. Detection and segmentation of structures can ease the image-text semantic alignments.

ers are used even in image classification (Dosovitskiy et al., 2020) and object detection (Xu et al., 2021).

The transformer architecture follows an encoder-decoder structure, though some applications may only use the encoder (Devlin et al., 2018) or the decoder (Brown et al., 2020) part. In the translation task, the encoder receives the input sentence of the language to be translated. The computed features go to the decoder, which outputs the correct translations. A natural way to adapt this architecture to an image captioning solution is feeding the decoder with image features, as depicted in Fig. 14.

After training, the described architecture should attend to specific parts of the image when writing a report. The performance of the network can be improved by easing the image-text semantic alignments that happen during training (Anderson et al., 2018; Li et al., 2020). For that, we can use the tooth instance segmentation techniques mentioned before. Further improvements may be achieved with other structures (restorations, prostheses, cysts, tumors). However, it is not always clear how effective they will be to alignment learning.

5 Achievements and Challenges

In dentistry, CNNs have been tested in photographs, radiographs, and tomographic images as auxiliary diagnostic tools (Shan et al., 2021). In digital photographs, CNNs have already been used to classify facial features, dental biofilms, oral cancer and detecting periodontal inflammations (Shan et al., 2021; Schwendicke et al., 2019). In bite-wings radiographs, CNN networks were used to detect carious lesions and segmented teeth (Lee et al., 2021; Shan et al., 2021). In periapical radiographs, they were used for classification, detection, and segmentation of teeth, as well as detecting caries, bone loss, and other dental pathologies (Shan et al., 2021; Schwendicke et al., 2019). In panoramic radiographs, the CNNs were used for tooth classification, root morphology, sex and age estimation, tooth segmentation and detection, periodontal bone loss, osteoporosis, atherosclerotic plaques in the carotid artery, and treatment prediction (Shan et al.,2021; Płotka et al., 2021; Krois et al., 2019; Silva et al.,2018; Silva et al., 2020; Jader et al., 2018; Bouchahma et al., 2019). In cephalometric radiographs, CNNs were used to detect anatomical landmarks (Shan et al., 2021; Płotka et al., 2021). Considering CBCT images, CNNs have been used to classify periapical lesions, classifying and detecting teeth and their structures, detecting vertical root fractures, anatomical points, cystic lesions and metastatic lymph nodes in patients with head and neck cancer, and segment the mandible (Hung et al., 2020; Shan et al., 2021; Schwendicke et al., 2019; Lee et al., 2020). In facial and intraoral scans, CNNs have been used for tooth segmentation (Płotka et al., 2021).

Despite the encouraging results showed by convolutional neural networks in the dentistry field, there are numerous other challenges to overcome:

- It is known that the efficiency of ML-based models depends on data quantity. Some peculiarities can make it difficult to gather a data set with an adequate size: The variability of human anatomy, conditions, and lesions contributes to the demand for an even larger amount of images to guarantee generalization performance; when considering a supervised approach, the difficulty resides on forming a suitable diversified data set of images annotated by experts or well-trained professionals; these two issues make the detection of a rare anomaly or lesion a complex task to tackle.

- It is noteworthy that the quality of the images also interferes in the annotation quality by making it difficult to see the labeled condition. For instance, two-dimensional images such as photographs can lose focus, and have inadequate lighting, and framing. Radiographs may present poor levels of brightness and contrast, overlapping structures, magnification, and distortions, in addition to not providing relevant three-dimensional information. On the other hand, 3D images have superior quality and have shown a tendency to replace 2D images, as equipment technologies evolve. However, these images are still not routinely used in clinical practice.

- Another issue to be evaluated is the heterogeneity of the dataset so that the model can be generalized and adopted in clinical tests. It is of underlying importance to gather a data set with images captured by different types of equipment and using distinct protocols, coming from different places, covering people with a wide age range, different ethnicities, and oral conditions (Tang, 2020). For a more accurate diagnosis, some studies suggest hybrid deep learning models that use radiographic images with diverse social-demographic and risk factors, as well as integrating clinical information or even combining knowledge-based algorithms and Machine Learning (Sarvamangala and Kulkarni, 2021; Shan et al., 2021).

- A further limitation to be controlled is the high rate of false alarms, as it may induce unnecessary treatments. For example, cases where the model tends to misidentify dental fractures or dental caries due to the difference in radiographic densities of these conditions.

- There are few public data sets available in the field. It would be interesting to develop a public dental data set repository, where researchers make their data set available, creating a robust volume of diversified and organized image databases. It would allow the development of algorithms with clinical practice application.

- Some ethical considerations should be made for the future incorporation of AI/ML-based systems into clinical practice. Issues such as responsibility for the privacy of patient data and for their diagnosis should be discussed to establish ethical norms for the clinical use of AI.

6 Concluding Remarks

Dentistry is a potential area to explore deep learning methods, and it still has many under-investigated gaps. Dental images are essential tools for dentists, from diagnosis to treatment follow-up. Throughout life, many of us are affected by prevalent dental conditions (e.g., caries, tooth loss, bone loss), and imaging tests, as well as other data (records, medications, treatments), are collected at various times. This dental history data is very rich for applying computer vision techniques, which can integrate and cross-reference such data to improve predictions.

References

Alex Krizhevsky, Ilya Sutskever and Geoffrey E. Hinton. (2012). Imagenet classification with deep convolutional neural networks. *Advances in Neural Information Processing Systems*, 25: 1097–1105.

Alexander Kirillov, Yuxin Wu, Kaiming He and Ross Girshick. (2020). Pointrend: Image segmentation as rendering. In *Proceedings of the IEEE/CVF Conference on Computer Vision and Pattern Recognition*, pp. 9799–9808.

Alexey Dosovitskiy, Lucas Beyer, Alexander Kolesnikov, Dirk Weissenborn, Xiaohua Zhai et al. (2020). An image is worth 16 × 16 words: Transformers for image recognition at scale. *arXiv preprint arXiv:2010.11929*.

Anil Gurses and Ayse Betul Oktay. (2020). Human identification with panoramic dental images using mask R-cnn and surf. In *2020 5th International Conference on Computer Science and Engineering (UBMK)*, pp. 232–237. IEEE.

Anselmo Garcia Cantu, Sascha Gehrung, Joachim Krois, Akhilanand Chaurasia, Jesus Gomez Rossi et al. (2020). Detecting caries lesions of different radiographic extension on bitewings using deep learning. *Journal of Dentistry*, 100: 103425.

Ashish Vaswani, Noam Shazeer, Niki Parmar, Jakob Uszkoreit, Llion Jones et al. (2017). Attention is all you need. In *Advances in Neural Information Processing Systems*, pp. 5998–6008.

Bernardo Silva, Laís Pinheiro, Luciano Oliveira and Matheus Pithon. (2020). A study on tooth segmentation and numbering using end-to-end deep neural networks. In *33rd SIBGRAPI Conference on Graphics, Patterns and Images (SIBGRAPI)*, pp. 164–171.

Bhornsawan Thanathornwong and Siriwan Suebnukarn. (2020). Automatic detection of periodontal compromised teeth in digital panoramic radiographs using faster regional convolutional neural networks. *Imaging Science in Dentistry*, 50(2): 169.

Brown, J. E. (2001). Advances in dental imaging primary dental care. *Journal of the Faculty of General Dental Practitioners*, 8(2): 59–62.

Christian Szegedy, Vincent Vanhoucke, Sergey Ioffe, Jon Shlens and Zbigniew Wojna. (2016). Rethinking the inception architecture for computer vision. In *Proceedings of the IEEE Conference on Computer Vision and Pattern Recognition*, pp. 2818–2826.

Christian Szegedy, Wei Liu, Yangqing Jia, Pierre Sermanet, Scott Reed et al. (2015). Going deeper with convolutions. In *Proceedings of the IEEE Conference on Computer Vision and Pattern Recognition*, pp. 1–9.

Dmitry V. Tuzoff, Lyudmila N. Tuzova, Michael M. Bornstein, Alexey S. Krasnov, Max A. Kharchenko et al. (2019). Tooth detection and numbering in panoramic radiographs using convolutional neural networks. *Dentomaxillofacial Radiology*, 48(4): 20180051.

Emre Avuçlu and Fatih Basçiftçi. (2020). The determination of age and gender by implementing new image processing methods and measurements to dental x-ray images. *Measurement*, 149: 106985.

Falk Schwendicke, Tatiana Golla, Martin Dreher and Joachim Krois. (2019). Convolutional neural networks for dental image diagnostics: A scoping review. *Journal of Dentistry*, 91: 103226.

Frank C. Setzer, Katherine J. Shi, Zhiyang Zhang, Hao Yan, Hyunsoo Yoon et al. (2020). Artificial intelligence for the computer-aided detection of periapical lesions in cone-beam computed tomographic images. *Journal of Endodontics*, 46(7): 987–993.

Gil Jader, Jefferson Fontineli, Marco Ruiz, Kalyf Abdalla, Matheus Pithon et al. (2018). Deep instance segmentation of teeth in panoramic x-ray images. In *31st SIBGRAPI Conference on Graphics, Patterns and Images (SIBGRAPI)*, pp. 400–407.

Gil Silva, Luciano Oliveira and Matheus Pithon. (2018). Automatic segmenting teeth in x-ray images: Trends, a novel data set, benchmarking and future perspectives. *Expert Systems with Applications*, 107: 15–31.

Hang Zhang, Chongruo Wu, Zhongyue Zhang, Yi Zhu, Haibin Lin et al. (2020). Resnest: Split-attention networks. *arXiv preprint arXiv:2004.08955*.

Hansang Lee, Minseok Park and Junmo Kim. (2017). Cephalometric landmark detection in dental x-ray images using convolutional neural networks. In *Medical Imaging 2017: Computer-Aided Diagnosis*, volume 10134, pp. 101341W. International Society for Optics and Photonics.

Hu Chen, Kailai Zhang, Peijun Lyu, Hong Li, Ludan Zhang et al. (2019). A deep learning approach to automatic teeth detection and numbering based on object detection in dental periapical films. *Scientific Reports*, 9(1): 1–11.

Hung, K., Yeung, A., Tanaka, R. and Bornstein. M. M. (2020). Current applications, opportunities, and limitations of AI for 3d imaging in dental research and practice. *International Journal of Environmental Research and Public Health*, 17(12).

Hyuk-Joon Chang, Sang-Jeong Lee, Tae-Hoon Yong, Nan-Young Shin, Bong-Geun Jang et al. (2020). Deep learning hybrid method to automatically diagnose periodontal bone loss and stage periodontitis. *Scientific Reports*, 10(1): 1–8.

Ian Goodfellow, Yoshua Bengio and Aaron Courville. (2016). Deep Learning. MIT Press.

IDATARESEARCH. https://idataresearch.com/how-many-dental-x-rays-are-performed-in-the-united-states/.

Jacob Devlin, Ming-Wei Chang, Kenton Lee and Kristina Toutanova. (2018). Bert: Pre-training of deep bidirectional transformers for language understanding. *arXiv preprint arXiv:1810.04805*.

Jayachandran, S. (2017). Digital imaging in dentistry: A review. *Contemp. Clin. Dent.*, 8(2): 193–194.

Jia Deng, Wei Dong, Richard Socher, Li-Jia Li, Kai Li et al. (2009). Imagenet: A large-scale hierarchical image database. In *2009 IEEE Conference on Computer Vision and Pattern Recognition*, pp. 248–255. IEEE.

Joachim Krois, Thomas Ekert, Leonie Meinhold, Tatiana Golla, Basel Kharbot et al. (2019). Deep learning for the radiographic detection of periodontal bone loss. *Scientific Reports*, 9(1): 8495.

Joel Jaskari, Jaakko Sahlsten, Jorma Järnstedt, Helena Mehtonen, Kalle Karhu et al. (2020). Deep learning method for mandibular canal segmentation in dental cone beam computed tomography volumes. *Scientific Reports*, 10(1): 1–8.

Jonathan Long, Evan Shelhamer and Trevor Darrell. (2015). Fully convolutional networks for semantic segmentation. In *Proceedings of the IEEE Conference on Computer Vision and Pattern Recognition*, pp. 3431–3440.

Joseph Redmon and Ali Farhadi. (2017). Yolo9000: Better, faster, stronger. In *Proceedings of the IEEE Conference on Computer Vision and Pattern Recognition*, pp. 7263–7271.

Joseph Redmon and Ali Farhadi. (2018). Yolov3: An incremental improvement. *arXiv preprint arXiv:1804.02767*.

Joseph Redmon, Santosh Divvala, Ross Girshick and Ali Farhadi. (2016). You only look once: Unified, real-time object detection. In *Proceedings of the IEEE Conference on Computer Vision and Pattern Recognition*, pp. 779–788.

Kai Chen, Jiangmiao Pang, Jiaqi Wang, Yu Xiong, Xiaoxiao Li et al. (2019). Hybrid task cascade for instance segmentation. In *Proceedings of the IEEE/CVF Conference on Computer Vision and Pattern Recognition*, pp. 4974–4983.

Kaiming He, Georgia Gkioxari, Piotr Dollár and Ross Girshick. (2017). Mask R-cnn. In *Proceedings of the IEEE International Conference on Computer Vision*, pp. 2961–2969.

Kaiming He, Xiangyu Zhang, Shaoqing Ren and Jian Sun. (2016). Deep residual learning for image recognition. In *Proceedings of the IEEE Conference on Computer Vision and Pattern Recognition*, pp. 770–778.

Karen Simonyan and Andrew Zisserman. (2014). Very deep convolutional networks for large-scale image recognition. *arXiv preprint arXiv:1409.1556*.

Kelvin Xu, Jimmy Ba, Ryan Kiros, Kyunghyun Cho, Aaron Courville et al. (2015). Show, attend and tell: Neural image caption generation with visual attention. In *International Conference on Machine Learning*, pp. 2048–2057. PMLR.

Ki-Sun Lee, Seok-Ki Jung, Jae-Jun Ryu, Sang-Wan Shin and Jinwook Choi. (2020). Evaluation of transfer learning with deep convolutional neural networks for screening osteoporosis in dental panoramic radiographs. *Journal of Clinical Medicine*, 9(2): 392.

Laís Pinheiro, Bernardo Silva, Brenda Sobrinho, Fernanda Lima, Patrícia Cury et al. (2021). Numbering permanent and deciduous teeth via deep instance segmentation in panoramic x-rays. In *17th International Symposium on Medical Information Processing and Analysis*, volume 12088, pp. 95–104. SPIE, 2021.

Lazar Kats, Marilena Vered, Ayelet Zlotogorski-Hurvitz and Itai Harpaz. (2019). Atherosclerotic carotid plaque on panoramic radiographs: Neural network detection. *Int. J. Comput. Dent.*, 22(2): 163–169.

Lee, J. H., Kim, D. H. and Jeong, S. N. (2020). Diagnosis of cystic lesions using panoramic and cone beam computed tomographic images based on deep learning neural network. *Oral Dis.*, 26(1): 152–158.

Liang-Chieh Chen, George Papandreou, Florian Schroff and Hartwig Adam. (2017). Rethinking atrous convolution for semantic image segmentation. *arXiv preprint arXiv:1706.05587*.

Liang-Chieh Chen, George Papandreou, Iasonas Kokkinos, Kevin Murphy and Alan L. Yuille. Deeplab: Semantic image segmentation with deep convolutional nets, atrous convolution, and fully connected crfs. *IEEE Transactions on Pattern Analysis and Machine Intelligence*, 40(4): 834–848.

Majed Bouchahma, Sana Ben Hammouda, Samia Kouki, Mouza Alshemaili and Khaled Samara. (2019). An automatic dental decay treatment prediction using a deep convolutional neural network on x-ray images. In *IEEE/ACS 16th International Conference on Computer Systems and Applications (AICCSA)*, pp. 1–4.

Mengde Xu, Zheng Zhang, Han Hu, Jianfeng Wang, Lijuan Wang et al. (2021). End-to-end semi-supervised object detection with soft teacher. *arXiv preprint arXiv:2106.09018*.

Mingxing Tan and Quoc Le. (2019). Efficientnet: Rethinking model scaling for convolutional neural networks. In *International Conference on Machine Learning*, pp. 6105–6114. PMLR.

Minyoung Chung, Jusang Lee, Sanguk Park, Minkyung Lee, Chae Eun Lee et al. (2021). Individual tooth detection and identification from dental panoramic x-ray images via point-wise localization and distance regularization. *Artificial Intelligence in Medicine*, 111: 101996.

Odeuk Kwon, Tae-Hoon Yong, Se-Ryong Kang, Jo-Eun Kim, Kyung-Hoe Huh et al. (2020). Automatic diagnosis for cysts and tumors of both jaws on panoramic radiographs using a deep convolution neural network. *Dentomaxillofacial Radiology*, 49(8): 20200185.

Olaf Ronneberger, Philipp Fischer and Thomas Brox. (2015). U-net: Convolutional networks for biomedical image segmentation. In *International Conference on Medical Image Computing and Computer-assisted Intervention*, pp. 234–241. Springer.

Oleksandr Boiko, Joni Hyttinen, Pauli Fält, Heli Jäsberg, Arash Mirhashemi et al. (2019). Deep learning for dental hyperspectral image analysis. In *Color and Imaging Conference*, volume 2019, pp. 295–299. Society for Imaging Science and Technology.

Oriol Vinyals, Alexander Toshev, Samy Bengio and Dumitru Erhan. (2015). Show and tell: A neural image caption generator. In *Proceedings of the IEEE Conference on Computer Vision and Pattern Recognition*, pp. 3156–3164.

Ozcan Karatas, Nazire Nurdan Cakir, Saban Suat Ozsariyildiz, Hatice Cansu Kis, Sezer Demirbuga et al. (2021). A deep learning approach to dental restoration classification from bitewing and periapical radiographs. *Quintessence International (Berlin, Germany: 1985)*, pp. 0–0.

Peter Anderson, Xiaodong He, Chris Buehler, Damien Teney, Mark Johnson et al. (2018). Bottom-up and top-down attention for image captioning and visual question answering. In *Proceedings of the IEEE Conference on Computer Vision and Pattern Recognition*, pp. 6077–6086.

Sabarudin, A. and Tiau, Y. J. (2013). Image quality assessment in panoramic dental radiography: A comparative study between conventional and digital systems. *Quantitative Imaging in Medicine and Surgery*, 3(1): 43–48.

Sarvamangala, D. R. and Kulkarni, R. V. (2021). Convolutional neural networks in medical image understanding: A survey. *Evol. Intel.*, 17.

Sepp Hochreiter and Jürgen Schmidhuber. (1997). Long short-term memory. *Neural Computation*, 9(8): 1735–1780.

Shah, N., Bansal, N. and Logani, A. (2014). Recent advances in imaging technologies in dentistry. *World Journal of Radiology*, 6(10): 794–807.

Shan, T., Tay, F. R. and Gu, L. (2021). Application of artificial intelligence in dentistry. *J. Dent. Res.*, 100(3): 232–244.

Shankeeth Vinayahalingam, Steven Kempers, Lorenzo Limon, Dionne Deibel, Thomas Maal et al. (2021). Classification of caries in third molars on panoramic radiographs using deep learning. *Scientific Reports*, 11(1): 1–7.

Shaoqing Ren, Kaiming He, Ross Girshick and Jian Sun. (2015). Faster r-cnn: Towards real-time object detection with region proposal networks. *Advances in Neural Information Processing Systems*, 28: 91–99.

Shinae Lee, Sang-il Oh, Junik Jo, Sumi Kang, Yooseok Shin et al. (2021). Deep learning for early dental caries detection in bitewing radiographs. *Scientific Reports*, 11(1): 16807.

Shu Liu, Lu Qi, Haifang Qin, Jianping Shi and Jiaya Jia. (2018). Path aggregation network for instance segmentation. In *Proceedings of the IEEE Conference on Computer Vision and Pattern Recognition*, pp. 8759–8768.

Szymon Płotka, Tomasz Włodarczyk, Ryszard Szczerba, Przemysław Rokita, Patrycja Bartkowska et al. (2021). Convolutional neural networks in orthodontics: A review.

Tang, X. (2020). The evolution and application of dental maxillofacial imaging modalities. *BJR—Open*, 2(1).

Tom B. Brown, Benjamin Mann, Nick Ryder, Melanie Subbiah, Jared Kaplan et al. (2020). Language models are few-shot learners. *arXiv preprint arXiv:2005.14165*.

White, S. C. and Pharoah, M. J. (2008). The evolution and application of dental maxillofacial imaging modalities. *Dental Clinics of North America*, 52(4): 689.

White, S. C. and Pharoah, M. J. (2009). *Oral Radiology: Principles and Interpretation*. Mosby/Elsevier.

WHO. https://www.who.int/health-topics/oral-health#tab=tab_1.

Wiwiek Poedjiastoeti and Siriwan Suebnukarn. (2018). Application of convolutional neural network in the diagnosis of jaw tumors. Healthcare Informatics Research, 24(3): 236–241.

Xiujun Li, Xi Yin, Chunyuan Li, Pengchuan Zhang, Xiaowei Hu et al. (2020). Oscar: Object-semantics aligned pre-training for vision-language tasks. In *European Conference on Computer Vision*, pp. 121–137. Springer.

Yann LeCun, Léon Bottou, Yoshua Bengio and Patrick Haffner. (1998). Gradient-based learning applied to document recognition. *Proceedings of the IEEE*, 86(11): 2278–2324.

Yu Song, Xu Qiao, Yutaro Iwamoto and Yen-wei Chen. (2020). Automatic cephalometric landmark detection on x-ray images using a deep-learning method. *Applied Sciences*, 10(7): 2547.

Yuma Miki, Chisako Muramatsu, Tatsuro Hayashi, Xiangrong Zhou, Takeshi Hara et al. (2017). Classification of teeth in cone-beam ct using deep convolutional neural network. Computers in Biology and Medicine, 80: 24–29.

Chapter 10

Malarial Parasite Detection in Blood Smear Microscopic Images: A Review on Deep Learning Approaches

Kinde Anlay Fante* and *Fetulhak Abdurahman*

ABSTRACT

Malaria is one of the deadly tropical and subtropical diseases in the world, especially in the developing countries of Africa, Asia, and Latin American continents. It is transmitted by the infected female Anopheles mosquito. Early diagnosis of malaria can reduce the mortality rate and its harmful consequences. The gold standard and widely used method of malaria diagnosis is the manual

---

Faculty of Electrical and Computer Engineering, JiT, Jimma University, Jimma, Ethiopia.
* Corresponding author: kinde.anlay@ju.edu.et

examination of blood smears using light microscopy. However, this method is subjective, time-consuming, and error-prone. The diagnosis result highly depends on the level of technical expertise and experience of the laboratory technicians. To improve the reliability and accuracy of this diagnosis method, automated computer-aided diagnosis (CADx) systems were proposed as a viable option. The CADx systems are used to detect the malarial parasites in the microscopic images and quantifying the level of infection. The methods proposed for detecting and classifying malarial parasites in blood smear microscopic images can be divided into two broad categories. The first methods category employs traditional image processing and classical machine learning algorithms. The second methods category employs deep learning methods for the detection and classification of malarial parasites. This work presents a comprehensive review of different methods for malarial parasite detection and classification in blood smear microscopic images. A methodological review of the recent deep learning techniques is given more emphasis. This review clearly shows the technical progresses attained in an attempt to solve this problem and future research directions.

1 Introduction

Malaria is one of the deadly tropical and subtropical diseases in many parts of the world, especially in developing countries of Africa, Asia, and Latin American continents. It is an infectious and noncontagious disease. It is transmitted by an infected female Anopheles mosquito, the carrier of the parasite (one of the Plasmodium species). There are five Plasmodium species known as Plasmodium malariae (P. malariae), Plasmodium knowlesi (P. knowlesi), Plasmodium ovali (P. ovali), Plasmodium vivax (P. vivax), and Plasmodium falciparum (P. falciparum). All the plasmodium species have different life stages in both human and mosquitoes, such as ring, trophozoite, schizont and gametocyte. The female Anopheles mosquito carries one of the Plasmodium species. Among these species, P. falciparum which is unicellular protozoa is the most deadly to human beings (Hanssen et al., 2010). A person infected with malaria shows symptoms of headache, fever, vomiting, and chill, which can be mild or severe. The parasite multiplies itself in the host liver and destroys the red blood cells (RBCs) of the host. The other different ways that malaria can be transmitted from one human to another include blood transfusion, from mother to fetus, and organ donation.

According to WHO's world malaria report, 228 million and 231 million cases of malaria occurred in 2018 and 2017, respectively. In 2018, 435,000 deaths were reported and 90% of the occurrences were from Africa (World Health Organization et al., 2019). Even though malaria has treatment drugs, early diagnosis and treatment significantly reduce its harmful consequences. Among malaria diagnosis methods the Rapid Diagnostic Test (RDT), polymerase reaction chain (PRC), and microscopy methods are widely used (Ojurongbe et al., 2013). The PCR works by detecting the antigens in the blood. It is 50 times more sensitive than microscopy. The cost of RDTs and PCR is very high. They require a high level of technical expertise, which is expensive. Microscopy has been widely used for more than 100 years for the diagnosis of malaria. It is a good method in terms of its suitability to characterize malarial parasite species, quantify the density of the parasite, and assess the effectiveness of the malaria treatment methods. The visual examination of thin and thick stained blood smears using a light microscope is the gold standard and a widely used method for malaria diagnosis (O'Meara et al., 2006). However, the manual method of examining stained blood smears using a light microscope is time consuming, error-prone, and subjective. The accuracy of diagnosis using this method highly depends on the expertise, experience and attentiveness of the laboratory technician during the procedure. In malaria-endemic and resource-constrained places, especially in rural portions of underdeveloped countries, the lack of well-trained laboratory technicians is an additional challenge in the efficient utilization of this method. Incorrect diagnosis leads to morbidity and socioeconomic problems in society. It can also lead to poor decision-making and planning in the malaria eradication process.

Malarial parasitic cells are heterogeneous in their form. The subjective nature of diagnosis comes primarily from the lack of standardization in the preparation of blood films, which results in staining variations of the slides and lighting differences between laboratories. Three types of staining protocols are widely used. These are Giemsa, Leishman and Wright staining protocols. In the manual microscopic examination of blood smears, the blood films can be prepared as thin or thick films. Thin blood film staining is used to detect the number of infected RBCs and the total number of RBCs (parasite density). Whereas, thick blood film staining is used to determine the number of parasites and white blood cells (WBCs) or to differentiate the parasite from the background. Giemsa stain is the most commonly used staining protocol whereas Leishman staining is suitable for the visualization of the morphology of white and red blood cells (Sathpathi et al., 2014).

It is imperative to use a reliable, low cost and fast computer-aided diagnosis (CADx) system to overcome the limitations of the manual examination of blood smears using a microscope. Recent advances in computer vision, particularly the introduction of deep learning techniques, have yielded promising achievements in detecting malarial parasites and, more broadly, for abnormalities detection in various modality medical images. A lot of studies have been conducted in the field of medical image analysis using different image processing techniques over the past few decades. The image processing techniques are used either for segmentation, classification, detection or an ensemble of them (for example, segmentation followed by classification). The image analysis techniques employed for automated malaria diagnosis can be categorized into three groups: (1) traditional image processing techniques using different characteristics of objects in the image such as histogram, shape, color, and texture to select the region of interests (ROIs) in an image, (2) classical machine learning techniques which use handcrafted features from an input image and use them to detect (classify) different object classes found in the image, (3) deep learning based techniques are those which automatically learn features from training image datasets and use them to classify (detect) different objects in an image.

Generally, the conventional image processing algorithms for the detection of target objects in an image have a poor performance (Poostchi et al., 2018; Park et al., 2016; Ross et al., 2006). This is due to the unreliability of the segmentation techniques when applied to images of a particular domain that are acquired in different environments. The classical machine learning-based approaches that use manually extracted features were also widely studied for classification and detection of abnormalities in medical image analysis (Rosado et al., 2016; Tek et al., 2010; Devi et al., 2018). The manual feature extraction and selection methods used by these techniques are sub-optimal. Moreover, these methods are computationally exhaustive for real-time applications.

Deep learning techniques for object detection and classification have recently gained popularity due to their capability to deal with the limitations of conventional image processing and machine learning techniques. Convolutional neural networks (CNNs) serve as the foundation for these deep learning algorithms. The main advantage of CNNs is their ability to automatically extract visual features from images. They can cope with the challenges of extracting robust and consistent features from images that are acquired in different environmental conditions which makes them suitable for automated malaria diagnosis from microscopic images of blood smears from different laboratories. Deep CNN architectures also achieve a good performance in detecting and classifying objects in images. Accordingly, the success of these networks was demonstrated in malaria parasite detection (Abdurahman et al., 2021), classification (Rahman et al., 2019) and subclassification (Hung et al., 2017) tasks.

The aim of this chapter is to review and analyze state-of-the-art deep learning-based approaches applied for the classification and detection of malarial parasites in microscopic images. A detailed methodological review of image processing and deep learning techniques for automated malarial parasite detection and classification in the blood smear images of microscopes is given. The analysis covers the traditional image processing algorithms, deep learning learning, and classical machine methods employed for malarial parasites classification and detection in microscopic images acquired using a camera attached to the eyepiece of a light microscope or digital microscope. Several deep learning techniques used for automated malaria diagnosis are compared in terms of their performance, the dataset used, and the network architectures. The existing challenges, perspectives, and future research directions of automated malaria diagnosis systems are also provided.

2 Traditional Image Processing and Classical Machine Learning Techniques

In this section, a review of automated malaria diagnosis systems that use conventional image processing and classical machine learning techniques is presented. Several studies have proposed various automatic malaria parasite detection techniques based on image processing and traditional machine learning methods in both thick and thin blood film microscopic images. As the oldest of the three approaches, the conventional image processing methods use different image processing techniques, such as thresholding followed by morphological operations or contouring to segment the parasite in microscopic images thick blood smears or malaria-infected red blood cells (RBCs) in the thin blood smear microscopic images. Simple rule-based classification systems are used to distinguish parasites from non-parasites or malaria-infected RBCs from those uninfected. In the case of classical machine learning approaches, manually designed feature extraction methods are used to extract distinct features of segmented malaria parasites or RBCs in microscopic images. Then, machine learning classification algorithms are used to classify the features as infected or non-infected RBCs in the microscopic images of thin blood films.

The authors of (Mas et al., 2015) applied temporal STD (standard deviation) to extract a single grayscale image from a video showing the movement of RBCs. After that, from the grayscale image a bi-modal histogram was computed to distinguish the pixels with high movement (high STD) from those with slow movement (low STD). A high STD indicates the presence of RBCs, whereas a low STD indicates low pixel activity, indicating the absence of RBCs. An automated end-to-end Plasmodium falciparum detection and quantification system was proposed in blood smear (thin film) microscopic images of mice and humans (Poostchi et al., 2018). In this work, a sequence of two methods was used to segment the RBCs. In this work, a sequence of two methods was used to segment the RBCs. The methods are multiscale detection using Laplacian of Gaussian (LoG) and active contours, specifically coupled edge profile active contours (C-EPAC). Then, to differentiate infected and uninfected RBCs, color and texture features were extracted. Nevertheless, the segmentation and feature extraction processes of these methods are computationally expensive. Furthermore, the number of images utilized to assess the performance of these algorithms is low. Hence, the reliability of these algorithms requires further assessment.

To reduce the computational complexity of image segmentation techniques, such as template matching, region growing, active contours, and morphological operations, the authors of (Moon et al., 2013) applied assumptions about the properties of isolated and clustered RBCs in microscopic images. Accordingly, they developed a malaria parasite detection algorithm based on the local maxima of the input image intensities composed of three channels, which are Mitotracker fluorescence (red), WGA-AlexaFlour488 fluorescence (green) and DAPI fluorescence (blue). This malaria parasites detection algorithm was also used to recognize the different stages of Plasmodium falciparum using image processing techniques without involving machine learning classifiers.

Several classical machine learning algorithms for detecting and classifying malarial parasites or infected RBCs are available in scientific literature. Among these algorithms, a binary class SVM (support vector machine) was employed to identify white blood cells (WBC) and P. falciparum trophozoites in stained thick blood film microscopic images, which were captured using a mobile phone camera (Rosado et al., 2016). In this work, an adaptive thresholding algorithm was first used to segment WBCs as well as trophozoite candidates in the image. Then, 314 features for every trophozoite candidate as well as 152 features for every WBC patch candidate were extracted. The extracted features were texture, color, and geometry descriptors. Finally, a binary class SVM was used to classify whether the segmented candidate is a WBC or not. Similarly, SVM was used to classify whether the segmented candidate regions are trophozoite or not. The algorithm has demonstrated 72.1% specificity and 98.2% sensitivity, for WBC detection. It has also achieved 93.8% specificity and 80.5% sensitivity for Plasmodium falciparum trophozoites detection.

The classification performance of k-nearest neighbor (KNN), linear discriminant classification (LDC), and linear regression (LR) for malaria diagnosis were compared in (Park et al., 2016). The classifiers used 23 morphological features obtained from phase images of unstained and live RBCs for malarial parasite classification. Similarly, Bayesian learning (BL) and support vector machine (SVM) classification algorithms were used to classify 96 textural and morphological descriptors which were extracted from different stages of P. falciparum and P. vivax infected erythrocytes (Das et al., 2013).

The performance of a KNN classifier for distinguishing four malarial species, namely P. malariae, P. ovale, P. vivax, and P. falciparum, in microscopic images of thin blood smears was evaluated in (Tek et al., 2010). The KNN classifier, with a relative distance metric and normalized L1 norm, used 83 input features which were extracted from color histograms, local area granulometry and shape measurements. The performance of the KNN classifier is reported to be better than FLD (fisher linear discriminant) and an artificial neural network. An ensemble classifier, which is constructed from SVM, k-NN, and ANN, was used to classify whether erythrocytes are infected or not in microscopic images of thin blood smears (Devi et al., 2018). The input features to the classifier are prediction error, chrominance channel histogram, LBP–GLCM, Gabor features, R–G channel difference histogram, Shanon entropy, saturation channel histogram, Havarda entropy, green channel histogram, Kapur entropy, Charvat entropy, and Renyi entropy. The comparison of the ensemble classifier with the individual classifiers (SVM, k-NN, and ANN) depicted that the ensemble classifier achieved the best performance.

A customized K-means clustering method used the value of pixel intensity as an input to detect malarial parasites (Purwar et al., 2011). However, the utilization of pixel intensity as the only feature to detect and identify malarial parasites without considering the textural and morphological information cannot be taken as a robust feature. A similar method was reported in (Nasir et al., 2012), which has used saturation, and the hue component of the Hue-Saturation-Intensity (HSI) color code as an input to a customized K-mean clustering method with the aim of segmenting malarial parasites, specifically the P. vivax gametocyte and trophozoite stages, in microscopic images.

A malaria diagnosis system with two feed-forward neural networks was proposed in (Ross et al., 2006). In this work, first, a histogram-based thresholding method was employed to segment the candidate regions for malarial parasites. Then, features were extracted from the candidate regions, including average eccentricity, average radius, shape, and color of erythrocytes. Finally, two separate models of feed-forward artificial neural networks, the first for classifying the segmented RBCs as infected or uninfected and the the other ANN model were used for classifying the life stage of the parasites for the infected RBCs. Unfortunately, the performance of threshold-based segmentation for P. malariae parasites detection is poor. Deep belief networks DBNs were also used to distinguish malarial parasites from other artifacts with gray level run length matrices, color coherence vectors, and color histograms as input features (Bibin et al., 2017).

We refer readers to review works in (Devi et al., 2019; Loddo et al., 2018; Poostchi et al., 2018; Sumi et al., 2021; Tek, 2009) for more information about traditional image processing and classical machine learning-based approaches for malarial parasites detection and classifications in blood smear microscopic images. Although the image processing algorithms and machine learning techniques have yielded promising performances for malarial parasite detection tasks, these methods are suboptimal. This is because the feature extractors used in these methods are designed manually. The processes of manual feature extraction selection and ranking are computationally very intensive and demand a high expertise level in particular application domains. Deep learning-based architectures, on the other hand, deal with the limitations of conventional hand-crafted feature extraction methods by learning inherently robust features from training data. The objective of this chapter is not to review the conventional malarial parasite diagnosis methods in detail but rather to present a detailed review of deep learning-based approaches for malaria diagnosis. As illustrated below, we further classify the deep learning-based malarial parasite detection approaches into patch-based, and end-to-end object detection approaches.

3 Deep Learning Approaches for Detection of Malarial Parasites

The deep learning approach is the recent trend in computer vision due to its remarkable performance in object detection as well as classification tasks of medical image analysis and other application domains. The main advantage of the deep learning approach in object detection and classification is that it does not require the design of a manual feature extractor, unlike the classical machine learning approaches. Instead, the optimal features are learned through training from data.

In this section, we present a review of deep learning techniques for detecting malarial parasites in blood smear microscopy images. All malarial parasites detection methods involve three basic operations. These are (1) selection of the malarial parasites candidate region (region proposal system), (2) extraction of features of the candidate regions (regions of interest), and (3) classification of the candidate regions as consisting of infected or healthy cells. In most of the deep learning methods, the feature extractor and classifier are integrated into the CNN architecture. The difference lies in how the candidate region selection algorithm (region proposal network) is integrated with the detection network. Based on these, the automated malarial parasite detection methods can be classified into three broad groups. The first group of algorithms uses CNN architectures to detect objects by extracting patches through conventional image possessing methods such as segmentation or via employing the sliding window method over the target image. Because of the large number of patches extracted from a single image and the computation of convolution for those patches these techniques have high computational complexity (Zhang, 2018). The second group (two-stage object detectors) uses a region proposal network (RPN) for the candidate region selection. Then the objects of interest in an image are identified by classifying the candidate regions. This group of deep learning models includes the Faster R-CNN (Ren et al., 2017), SPP-Net (Kaiming et al., 2014), Fast R-CNN (Girshick, 2015), and R-CNN (Girshick, 2014) network architectures. Since they have a separate proposed region and classification unit, these algorithms are referred to as two-stage object detection models. The last group of object detection models is known as a single-stage object detection model which includes SSD model (Liu et al., 2016) and the YOLO decedents (V1 (Redmon et al., 2016), V2 (Redmon and Farhadi, 2016), V3 (Redmon and Farhadi, 2018), and V4 (Bochkovskiy et al., 2020)). Rather than proposing regions of interest separately, this group of techniques treats the object detection problem as a single regression problem. In the following sections, we present a detailed review of deep learning techniques which were applied for either classification or detection of malarial parasites in microscopic images of thin or thick stained blood films.

3.1 Patch-Based Approach of Malarial Parasites Detection using Deep CNN Models

The diagnosis of malaria using the microscopy method requires the identification of the parasites in the blood sample and evaluating the degree of infection or number of parasites. The automated computer-aided diagnosis (CADx) systems must detect the parasites in the blood smear images and count the number of parasites in them. Hence, the detection of parasites is the crucial step of automated malaria diagnosis systems. The naive approach to detecting malarial parasites involves designing separate algorithms for candidate region selection and the candidate region classification into parasited or uninfected cells. The candidate region (region of interest) selection is mainly achieved using segmentation algorithms or the sliding window method. Then the region of interest is classified as parasited or healthy using convolutional neural networks (CNNs). The general block diagram of the patch-based approach for malarial parasites detection is shown in Fig. 1.

Figure 1: Block diagram patch-based technique for detecting malarial parasites using CNN.

As shown in Fig. 1, the region of interest selection involves traditional image processing methods such as color space conversion and stain normalization. The image segmentation method separates the cells from the background. The post-processing step involves the removal of tiny segmented objects, cropping the candidate region from the preprocessed image, and resizing the patch to fit into the input layer of the CNN. The CNN architecture extracts the features using convolution layers and classifies the patch using fully connected layers. Most of the existing deep learning-based approaches for malarial parasites' detection are patch-based, where the major task in segmenting the parasites' location is based on traditional image processing techniques. For

Table 1: Summary of the existing deep learning techniques for malarial parasites detection.

Reference	Dataset	Species	Staining	Blood Smear Type	Category	Method	Performance
Yang et al., 2019	1819 images	P. falciparum	Giemsa	Thick	classification	CNN	Sensitivity 92.59%, Specificity 94.33%
Yu et al., 2020	2967 cell images	P. falciparum	Giemsa	Thick	classification	Custom CNN	Sensitivity 98.1%, Specificity 99.2%
Liang et al., 2016	27,578 Erythrocyte images	P. falciparum	Giemsa	Thin	classification	CNN	Sensitivity 96.99%, Specificity 97.75%
Mehanian et al., 2017	classification	P. falciparum	Giemsa	Thick	classification	CNN	Sensitivity 91.6%, Specificity 94.1%
Rajaraman et al., 2018	27,558 cell images	P. falciparum	Giemsa	Thin	classification	pre-trained CNN	Sensitivity 98.1%, Specificity 99.2%
Rahman et al., 2019	27,558 cell images	P. falciparum	Giemsa	Thin	classification	CNN, VGG16 CNNEx-SVM	Custom acc. 95.97%, VGG16 acc. 97.64%, CNNExSVM, 94.77%
Umer et al., 2020	27,558 cell images	P. falciparum	Giemsa	Thin	classification	CNN	Accuracy 99.96%
Hung et al., 2017	1300 images	All species	Giemsa	Thin	detection	Faster R-CNN	98% of accuracy
Yang et al., 2019	2567 images	P. vivax	Giemsa	Thick	detection	cascaded YOLOV-2	mAP of 79.22%
Chibuta and Acar, 2020	1182 images	P. falciparum	Leishman	Thin	detection	modified YOLOV-3	mAP of 90.2%
Abdurahman et al., 2021	1182 images	P. falciparum	Leishman	Thick	detection	modified YOLOV-3, YOLOV-4	YOLOV3, mAP of 96.32%, YOLOV-4 mAP of 96.14%
Rahman et al., 2021	65,645 cell images	P. falciparum	Giemsa	Thin	Classification	VGG-16, VGG-19, Xception, ResNet-50, GoogleNet, custom CNN	best accuracy 99.35, pre-trained VGG-19
Nakasi et al., 2020	643 images	P. falciparum	Giemsa	Thick	Detection	pre-trained faster R-CNN SSD RetinaNet	best mAP 0.9448, pre-trained Faster R-CNN
Kumar et al., 2021	27,558 cell images	P. falciparum	Giemsa	Thin	classification	Alex-Net Mosquito-Net VGG-16 ResNet-50 DenseNet-121 Xception-Net	best accuracy 96.6%, Mosquito-Net
Peñas et al., 2018	363 images	P. falciparum, P. Vivax	Giemsa	Thin	classification	Inceptionv3	Sensitivity 98.62%, Specificity 96.82%
Loh et al., 2021	297 cell images	P. falciparum	Giemsa	Thin	Detection	Mask R-CNN	mAP of 0.731
Elangovan and Nath, 2021	27,558 cell images	P. falciparum	Giemsa	Thin	classification	Alexnet VGG-19 VGG-16 Resnet-50 Resnet-18 Squeezenet Resnet-101 Inception-v3 Mobilenet-v2 Xception Googlenet Densenet-201	Sensitivity 97.9%, Specificity 97.8%
Rahman et al., 2021	63,645 cell images	P. falciparum, P. vivax	Giemsa	Thin	classification	VGG-16 VGG-19 Resnet-50 Googlenet Xception	Best model: Sensitivity 96.32%, Specificity 97.41%
Kassim et al., 2020	965 cell images	P. falciparum	Giemsa	Thin	Segmentation and Detection	RBCNet	F1 score 97.94%, Recall 98.39%

instance, in (Umer et al., 2020), a CNN architecture with five convolutional layers and three fully connected layers is used. The regional proposal method employs traditional image processing methods in this group of methods, i.e., the region of interest (ROI) is extracted using image processing algorithms such as segmentation or via the sliding window method from an input image. Then, the extracted region is classified using a CNN-based classifier. The number of patches generated per image by these algorithms is relatively high. The computational complexity of convolutional operations on a large number of patches is prohibitive for real-time applications. Following the introduction of LeNet (LeCun et al., 1998) and the outstanding performance of AlexNet (Krizhevsky et al., 2012) at the ILSVRC (Russakovsky et al., 2015), these algorithms were used for the detection and classification of malarial parasites. A detailed summary of the deep learning-based studies on automated malaria diagnosis is depicted in Table 1.

Three deep learning CNN models (LeNet-5, AlexNet, and GoogLeNet) were used by authors of (Dong et al., 2017) for the classification of malarial parasites in which thresholding followed by the application of the Hough Circle transform (Duda and Hart, 1972) for the extraction of ROIs. Among the compared models in this work, GoogLeNet achieved the highest performance with an accuracy of 98.13%. CNN based pre-trained and customized models were proposed to screen malaria cases in thin blood film microscopic images (Rajaraman et al., 2018). In this study, the authors used models of pre-trained CNNs including Xception, AlexNet, ResNet-50, DenseNet-121 and VGG-16 as feature extractors by taking the fully connected layers before the classification layer and optimizing the learning rate, utilizing Nesterov's SGD and L2-regularization hyper-parameters using a randomized grid search algorithm. Statistical analysis was performed using the Kruskal-Wallis H test (Vargha and Delaney, 1998) and post-hoc analysis (Kucuk et al., 2016) to select the best-performing model. In comparison to the other models, the authors claimed that the pre-trained ResNet 50 performed the best in distinguishing infected erythrocytes.

A local intensity minima-based patch extraction (ROI candidate selection) method from a video was used in (Gopakumar et al., 2018). This video was made up of a multiple Field-of-Views (FoV) focus stack of Leishman stained microscopic images for recognition of Plasmodium falciparum and the life-cycle stages of malarial parasites. The extracted focus stack of patches were classified using a customized CNN model with specificity and sensitivity of 98.50% and 97.06%, respectively. Another customized CNN-based classifier model was used to differentiate the infected erythrocytes from healthy ones in microscopic images of stained thin blood smears by the authors of (Liang et al., 2016). The authors of (Torres et al., 2018) reported the development of Autoscope, which is a prototype digital device, using the CNN model for identifying and quantifying malarial parasites in Giemsa stained microscopic thick blood film images. A pre-trained VGG19 model, which is a deep CNN model, whose last three fully connected layers were replaced by a SVM classifier was used for malarial parasites classification in (Vijayalakshmi and Kanna, 2019). Similarly, the study in (Yang et al., 2019) explored the possibility of applying CNN models for identification of malarial parasites in microscopic images of thick blood films. The candidate malarial parasites location selection was carried out using a global minimum screening algorithm iterative method and a mobile application was also developed from the image pixel intensity values exhibiting a promising performance of 97.34% of area under curve AUC and 97.26% of accuracy at the patch level. It has also shown a good localization accuracy for detected parasites at the image level (Vijayalakshmi and Kanna, 2019).

In (Quinn et al., 2014), two image classification models were proposed for the detection of malarial parasites. The first classifier was used to extract the connected components in the image and moment feature descriptors from patches which are selected by applying the sliding window method and then it classifies them using extremely randomized trees. The second method involves using a customized CNN model to distinguish between infected and healthy RBCs in microscopic images of thick blood smears. With 97 percent average precision, the CNN-based classifier model has outperformed the decision tree classifier (Vijayalakshmi and Kanna, 2019). Three CNN-based classifier models for classifying malarial parasites in segmented microscopic images of thin blood smears were also presented in (Rahman et al., 2019). The first is a customized CNN model, which has 19 layers. The second model involves a transfer learning approach, specifically a pre-trained VGG16 model, which was obtained by training the original model by freezing layers from one to 16. The third model used a pre-trained VGG16 model as a feature extractor and SVM as a classifier to detect the malarial parasites. The performance of the ensemble of these three models was also investigated in this work. The second model (VGG16 model with transfer learning) was reported to be the best performing model for classifying malarial parasites.

The authors of reference (Pan et al., 2018) presented another deep convolutional neural network model based on the architecture LeNet for the detection and classification of malarial parasites in microscopic thin blood smear images. The authors applied an image interpolation algorithm in the spatial feature domain as a data augmentation technique to overcome the over-fitting problem. A discussion on the benefits of visualizing image features extracted by the convolution and pooling layers for understanding the visual scene learning strategy by the deep learning models in the malarial parasites classification problem can be found in (Sivaramakrishnan et al., 2017). The study in (Delahunt et al., 2019) attempted to address the major gap in the previous machine learning models by proposing a patient-level malarial parasites detection (training and validation) method. Two customized CNNs were applied to detect two stages of malarial parasites (ring and late-stage). A smart phone software application for malaria diagnosis in microscopic images of thick blood films that employed a patch-based customized CNN algorithm was reported in (Yang et al., 2019). In their work, the authors applied the histogram analysis of the image intensity values in order to select the candidate malarial parasites locations in the image. From their experimental results, they reported a good performance at both the patch and image levels. Generally, these patch-based methods are the most computationally complex among the deep learning approaches for object detection in images.

3.2 Two-Stage Malarial Parasite Detection Networks

The patch-based malarial parasite detection algorithms have two drawbacks. First, in the algorithms that use segmentation methods to select candidate ROIs, the accuracy of the segmentation methods is unreliable. Hence, these malaria parasites detection algorithms are unreliable. Second, the other shortcoming of the patch-based algorithms is the generation of a large number of patches per image in the sliding window methods which makes their computational complexity too high for real-time applications.

The two-stage object detection approaches have attempted to deal with the drawbacks of patch-based methods using region proposal algorithms (deep learning models for selection of candidate regions through training). For example, R-CNN (Girshick et al., 2014) resolves the drawbacks of the sliding window technique by generating a fixed number of ROIs (2k region proposals) using a selective search algorithm (Uijlings et al., 2013). The generated ROIs are passed to pre-trained CNN architectures such as VGG (Simonyan and Zisserman, 2015) for feature extraction. The extracted features by the CNN model are passed to category-specific linear SVMs for classification and bounding box regressors for localization. Despite its improvements over the conventional sliding window approach, R-CNN is relatively slow due to the overlapping CNN computations for each ROI, and the complex training process for three different models. The other architectures of this group include Fast R-CNN and Faster R-CNN. Fast R-CNN (Girshick, 2015) generates the region proposals from a shared feature map rather than passing every region proposal through the pre-trained CNN for feature extraction. The other improvement in the Fast R-CNN architecture has been the ROI pooling layer that generates fixed-sized ROIs, an input to the fully connected layers. Then, the feature outputs from the fully connected layers are passed to the classification and localization networks. A selective search algorithm is used in both R-CNN and Fast R-CNN architectures which are computationally expensive. These models also have a relatively low speed and accuracy compared to single-stage detection models. Faster-RCNN, with a more advanced and efficient approach for generating region proposals, enhances the object detection performance of R-CNN and Fast-RCNN. The detailed architectural illustration of the Faster R-CNN is given below.

3.2.1 Faster R-CNN

In Faster R-CNN (Ren et al., 2017), the selective search algorithms used in Fast R-CNN are replaced by Region Proposal Network (RPN). RPN is a cost-free end-to-end, a fully convolutional network trained to generate region proposals from an input image. Figure 2 depicts the architecture of the RPN in Faster R-CNN models. During the training phase, the RPN learns the region proposal

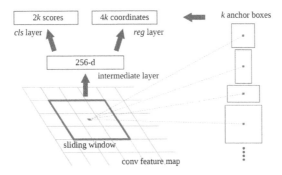

Figure 2: Architecture of RPN (Ren et al., 2017).

using the anchor box concept, which is labeled as positive (having object) or negative (background) based on their intersection over the union (IOU) value with the ground truth objects. In addition, the RPN uses a multi-task loss function, which is defined as:

$$L(\{p_i\}, \{t_i\}) = \frac{1}{N_{cls}} \sum_i L_{cls}(p_i, p_i^*) + \lambda \frac{1}{N_{reg}} \sum_i p_i^* L_{reg}(t_i, t_i^*) \quad (1)$$

where p_i is the predicted probability of anchor i, p_i^* is the ground truth label of anchor i (0: for background, 1: for object), t_i represent vectors of coordinates for predicted bounding boxes, t_i^* represents vectors of coordinates for positive anchor boxes, L_{cls} is the classification binary log loss, L_{reg} is the regression smoothed L1 loss, N_{cls} and N_{reg} are normalization factors and λ is the balancing weight. The complete architecture of Faster R-CNN is depicted in Fig. 3.

A few studies have used the two-stage deep learning object detection models for malarial parasites' detection. In (Hung et al., 2017), for example, a pre-trained Faster RCNN (Region-based CNN) model was employed to dis-

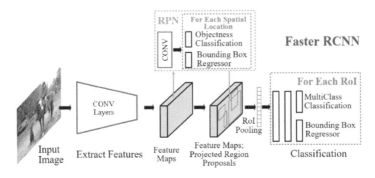

Figure 3: Architecture of Faster R-CNN (Liu et al., 2018).

tinguish RBCs from non-RBC objects. Then AlexNet was applied after detecting non-RBC objects, including leukocytes and malaria-infected RBCs, to further classify the detected infected RBCs into one of the categories of gametocyte leukocyte schizont ring and trophozoite. In comparison to the baseline model that employed conventional manual feature extraction, the Faster RCNN achieved a better performance. In (Loh et al., 2021), a Mask-RCNN based two-stage semantic segmentation approach was proposed to detect uninfected reticulocyte, ring, trophozoite, and schizont classes from microscopic images of cultured P. falciparum parasites using human RBCs cells. Another study in (Kassim et al., 2020) proposed a cascade of two deep learning models: RBC cell clustering-based segmentation from a thin smeared microscopic image and a two-stage Faster RCNN detector to predict individual RBCs infected by P. falciparum as the final output.

3.3 Single-Stage Malarial Parasite Detection Networks

Despite the success of two-stage and single-stage object detection deep learning models in various computer vision applications, including medical image analysis, their application for detecting malarial parasites has not been explored extensively. The one-stage deep learning-based models for object detection are computationally less expensive than the two-stage detectors. Moreover, the region proposal network and the detection head being separate, trained separately with long computation times make the two-stage object detection models inappropriate for real-time application. Due to such drawbacks, one-stage detectors such as SSD (Liu et al., 2016) and YOLO (Redmon and Farhadi, 2018) were proposed. These algorithms avoid using the region proposal network, enabling them to achieve an improved detection speed while maintaining acceptable detection accuracy. Different scholars proposed numerous one-stage object detectors. The selection of a deep learning model depends on different metrics for a given application domain. Some models are selected for their high detection performance, while others are selected for their low computation overheads. The most commonly used one-stage object detection models are SSD, YOLO, and RetinaNet in terms of their performance, while the MobileNet versions are selected with a less computational overhead for low-end devices such as mobile phones. In the following sections, we discuss the detailed architectural descriptions of the SSD and YOLO models.

3.3.1 Single Shot MultiBox Detector (SSD)

SSD takes the classification and localization steps in object detection as a single forward pass of the convolution network. SSD is a fully convolutional network having backbone feature extractor networks such as VGG16 (Simonyan

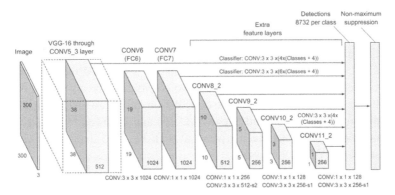

Figure 4: Architecture of SSD model (Liu et al., 2016).

and Zisserman, 2015) followed by additional six auxiliary convolutional feature layers, which decrease the size of the feature map generated from the backbone network. Each of the added feature maps is responsible for detecting objects, where the shallow layers detect small sized objects while the deeper layers detect larger objects. In addition, SSD takes advantage of a set of predefined anchor boxes (prior or default boxes) with different aspect ratios and scales to handle objects of different sizes and shapes. The SSD architecture is shown in Fig. 4. Each element of the detection feature map contains multi-scale predefined anchor boxes, which are either positive (matched) or negative (unmatched) boxes based on their intersection over union (IOU) ratio with the ground truth objects.

SSD computes offsets for predefined anchor box coordinates using smooth L_1 loss and their associated class confidence score using categorical cross-entropy loss during training. The final outputs of the SSD detection models are obtained by performing non-maximum suppression (NMS) on multi-scale refined bounding boxes.

3.3.2 You Only Look Once Version 3 (YOLOV3)

YOLOV3 (Redmon and Farhadi, 2018) is a potent single-stage model for object detection in images which has been used in a variety of applications including malarial parasites' detection in microscopic images. It outperforms its predecessors YOLOV1 (Redmon et al., 2016) and YOLOV2 (Redmon and Farhadi, 2016) in object detection accuracy, localization, and detection speed. It utilizes one CNN for predicting the class probabilities of candidate objects and their corresponding coordinates of bounding boxes in an input image. In addition, it has benefited from the utilization of novel features such as Feature Pyramid Networks FPN (Lin et al., 2016) for the purpose of multi-scale prediction and Darknet53, a new backbone network that employs residual networks, for the extraction of robust features.

The structure of the YOLOV3 network is constructed from the backbone CNN and object detection heads. It utilizes a Darknet53 which has a single convolution layer with a size of 3 x 3 and 23 residual units or five Resnet Blocks of convolution layers with 1 x 1 and 3 x 3 sizes. At the start of each Resnet block, there are five strides and two convolution layers in order to reduce the size of an input image. The convolution layers are preceded by batch normalization (BN) (Ioffe and Szegedy, 2015) and followed by an activation function known as Leaky ReLU (Maas, 2013). Figure 5 depicts the detailed structure of the YOLOV3 model.

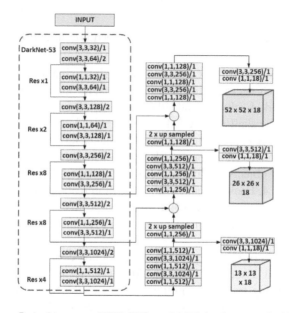

Figure 5: Architecture of YOLOV3 model (Abdurahman et al., 2021).

The YOLOV3 model has detection heads known usually as yolo layers. In this model, the input image is divided into different $S \times S$ grid sizes by these detection heads. The task of each grid cell is to detect objects whose center lies in that particular grid cell. Specifically, in detection headers, grid cells are assigned to anchor boxes whose centers are the centers of the grid cells. Object bounding boxes with different scale and aspect ratios are obtained using the K-means clustering method from the information of the object ground truth. In order to obtain features maps for efficient detection of different sizes of objects in an image, anchor boxes with large medium, and small scales are signed to small, medium, and large feature maps, respectively.

For detecting objects of various sizes, YOLOV3 extracts feature maps that are rich in both semantic and spatial information with different scales with the help of a modified FPN. The Darknet53 structure generates the first feature map as its final output, An additional convolution operation using filter sizes of $1\times 1, 3\times 3, 1\times 1, 3\times 3, 1\times 1$, and 1×1 are performed on the first feature map, and the outputs of the convolution operations are upsampled by two to form the second feature map. Then the concatenation of the upsampled feature maps and the feature maps created by the earlier layer is performed. The generation of the third feature scale is carried out by convolving the second feature scale with the above kernel sizes and upsampling the outputs of the convolution operation by two.

For a 416×416 input image, the feature map sizes in the YOLOV3 object detection model are 13×13 (first detection layer), 26×26 (second detection layer) and 52×52 (third detection layer). After these three feature maps are generated, YOLOV3 refines the features by performing six (1×1 and 3×3) convolution operations prior to the detection layer computations. The model generates a $N \times N(B \times 4 + 1 + C)$ feature tensor at all the three aforementioned detection layers as depicted in Fig. 6. Here, the detection layer feature map's grid size is indicated by $N \times N$. Where B denotes quantity of bounding boxes (three in YOLOV3) which must be predicted for every grid cell. For every bounding box, four values of the offset coordinate (center point, width and height) denoted by t_x, t_y, t_w, t_h and their corresponding confidence scores are computed. The confidence score is an indication of the probability of the existence of an object in an image, $P_r(Object)$. Its value lies between 0 and 1, where 0 indicates the absence of an object, and 1 indicates the presence of an object in an image. The confidence of precision (IOU_{Pred}^{Truth}) describes the intersection over union ratio (IOU) between the ground truth and predicted bounding boxes of an object in an image. C denotes the conditional probabilities of the classes ($P_r(Class_i|Object)$) of the target object bounding boxes, which are predicted using the grid cells. In (Abdurahman et al., 2021), a C value has been chosen equal to 1 since the proposed method detects one species of the malarial parasites, which is Plasmodium falciparum, in the microscopic image. The prediction of the four offset values in YOLOV3 is carried out relative to the particular grid cell contrary to the bounding boxes' actual target object dimensions in an image. Assuming that the grid cell center, which predicts the bounding box of an object, is at the offset of (C_x, C_y) from the target image in the top left corner and an anchor box with width (p_w) and height (p_h) is assigned, then the actual dimensions of the bounding box (b_x, b_y, b_w, b_h) are computed using the following equations:

$$\begin{aligned} b_x &= \sigma(t_x) + C_x \\ b_y &= \sigma(t_y) + C_y \\ b_w &= p_w \exp(t_w) \\ b_h &= p_h \exp(t_h) \end{aligned} \quad (2)$$

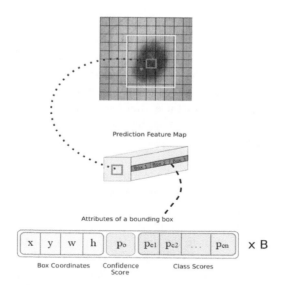

Figure 6: YOLOV3 prediction output tensor shape.

where, σ denotes the sigmoid function that is applied to place the centers of predicted bounding boxes within the grid cells with b_x and b_y denoting the centers of the bounding boxes and anchor boxes of width (b_w) and height (b_h).

The loss function used during the training phase of the model is binary cross-entropy which computes the class probabilities and confidence scores. The mean squared error loss function computes the loss in predicting coordinates of the bounding boxes of the target objects in an image. The localization and classification errors of the anchor boxes with the responsibility of detecting target objects are also computed. YOLOV3 uses a loss function that has three components, as given below.

Total loss = $Loss_{loc} + Loss_{obj} + Loss_{cls}$

$$\begin{aligned}
&= \lambda_{coord} \sum_{i=0}^{S^2} \sum_{j=0}^{B} 1_{ij}^{obj} \left[(x_i - \hat{x}_i)^2 + (y_i - \hat{y}_i)^2 \right] \\
&+ \lambda_{coord} \sum_{i=0}^{S^2} \sum_{j=0}^{B} 1_{ij}^{obj} \left[(\sqrt{w_i} - \sqrt{\hat{w}_i})^2 + (\sqrt{h_i} - \sqrt{\hat{h}_i})^2 \right] \\
&+ \sum_{i=0}^{S^2} \sum_{j=0}^{B} 1_{ij}^{obj} l(C_i - \hat{C}_i) + \lambda_{noobj} \sum_{i=0}^{S^2} \sum_{j=0}^{B} 1_{ij}^{noobj} l(C_i - \hat{C}_i) \\
&+ \sum_{i=0}^{S^2} 1_i^{obj} \sum_{c \, \varepsilon \, classes} l(p_i(c) - \hat{p}_i(c))
\end{aligned} \quad (3)$$

where the 1_{ij}^{obj} denotes a predicted object bounding box of the ith grid cell which uses the jth anchor box. 1_i^{obj} denotes the presence of a target object in the ith grid cell. The optimization parameters λ_{coord} and λ_{noobj} are used to accentuate the localization error and decrease the confidence score for anchor boxes of unsuccessful object detection attempts, respectively. The detection feature map's grid size is represented by S and B number of predefined bounding boxes are assigned to each grid cell. The four coordinates of the estimated bounding box are denoted by $(\hat{x}_i, \hat{y}_i, \hat{w}_i, \hat{h}_i)$ and the coordinate of the anchor box for the ground truth are denoted by (x_i, y_i, w_i, h_i). $\hat{C}_i$ represents the confidence score of the prediction, whereas C_i denotes the value of the correct confidence score. The prediction of the class probabilities of anchor boxes are denoted by the $\hat{p}_i(c)$, while the correct class probabilities for the target object ground truth are denoted by $p_i(c)$.

3.3.3 You Only Look Once Version 4 (YOLOV4)

YOLOV4 (Bochkovskiy, 2020) has a superior performance when compared to other deep learning models such as EfficientDet (Tan et al., 2019). The performance of this model in terms of object detection accuracy and speed for real-time applications has been enhanced by combining several features which have been proposed by other studies in the model structure with an insignificant inference time computational cost. The methods added to the YOLOV4 structure are divided into two categories. These are known as BOS (Bag of Specials) and BOF (Bag of Freebies). The two methods are used in both the backbone as well as detector modules. In the BOS part of this model, the backbone network is highly coupled with SSP. This coupling is used to enhance the feature maps of the detection layers by improving the size of the receptive fields of these layers. Cross-Stage-Partial DarkNet53 (CSPDarknet53) (Liu et al., 2018) is used as a backbone of YOLOV4. The main idea in this network structure is the introduction of a special connection known as cross-stage-partial (CSP). This forms a cross-layer connection in both the backbone and detection layers to enhance the robustness of the feature extraction stage with an insignificant computational cost. In contrast to YOLOV3, which uses FPN, YOLOV4 uses a modified version of PANet (Maas, 2013). The high-level and low-level feature maps are concatenated to enrich the geometric and semantic information of the target object using this PANet architecture. It uses a new activation function known as Mish, unlike YOLOV3, which uses Leaky ReLu, in its BOS part, which has a lower computational cost. The BOF part mainly involves data augmentation methods such as SAT (self adversarial training), Mosaic, and CutMix. It also uses Cross Mini-Batch Normalization (CmBN) to normalize batches of training images, and Drop Block regularization is used as a drop-out unit to prevent overfitting problems. Another freebie in the YOLOV4 model is the Complete IOU CIOU loss function that computes the level of overlap between the predicted bounding box and ground truth bounding box of a target object in an image.

YOLOV4 detection heads are similar to that of YOLOV3 in that both have three (3) detection feature maps. The feature maps are generated by concatenating feature maps from different layers of the convolution. The YOLOV4 model has 161 layers which are significantly higher than the YOLOV3 model with has 106 network layers. However, its performance was evaluated on the MS COCO dataset. Figure 7 shows the network architecture of YOLOV4.

A modified YOLOv3 architecture has also been proposed in (Chibuta and Acar, 2020) for detecting malarial parasites in microscopic images. This modified model has been obtained using a residual bottleneck network in its feature extractor network and reducing the number of convolution layers of the original detection network of the YOLOv3 architecture. The modified YOLOv3 architecture was used to detect Plasmodium falciparum parasites in stained thick blood film microscopic images acquired by a digital microscope camera and a smartphone camera. This architecture has a lower computational cost and higher inference speed than the original YOLOV3 version, but it has a lower detection accuracy. An automated malarial parasites, particularly *Plasmodium vivax*, detection system in microscopic stained thin blood film images was proposed in (Yang et al., 2019). The cascade of two network models, AlexNet classifier with transfer learning followed by YOLOv2, were also utilized to reduce the model's false positive prediction of the YOLOV3 model. The cascaded model outperformed the original model of YOLOv2 in terms of accuracy.

In (Abdurahman et al., 2021), two network architectures of YOLOV3 were presented. In the YOLOV3 model, the introduction of feature pyramid networks (FPN) enables it to detect objects of different sizes through its multi-scale feature map model. The FPN structure enables the concatenation of deep level features and shallow level features which have the same feature size. The deeper features of CNN have large receptive fields which are used to detect large objects. In single-stage object detection models, the feature extracted from shallow layers possess less semantic information and more primitive features (geometric information), but their small receptive fields are suitable to detect small objects in an image. The FPN structure is used to merge deep features which are rich in semantic information with shallow level features that are rich in geometric information. As the features extracted using the backbone network get deeper, the extracted semantic and geometric descriptors pertinent to small-sized object detection, such as malarial parasites in microscopic images, are lost. The three layers of the detection network have feature map scales of 13×13, 26×26, and 52×52 in the original model of YOLOV3. It has been revealed that the size of receptive fields of the above feature maps

Malarial Parasite Detection in Blood Smear Microscopic Images ■ 217

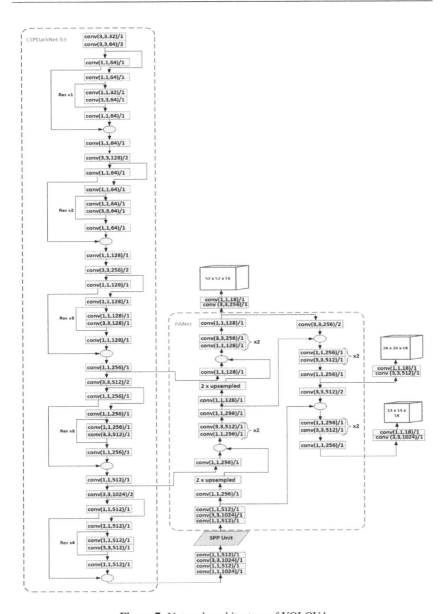

Figure 7: Network architecture of YOLOV4.

is larger than the size of P. falciparum in microscopic images. One of the possible modifications to this model was obtained by adding one additional layer in the detection network to the existing detection layer of the original model of YOLOV3, resulting in a four detection layers for the modified architecture. This added layer has been found to improve the performance of the modified

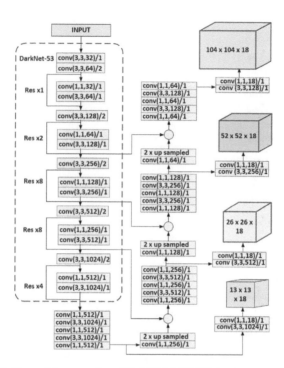

Figure 8: The architecture of modified YOLOV3 (Abdurahman et al., 2021).

model of YOLOV3 for the detection of small-sized objects. This is due to the that the concatenation of the features from the deep and shallow layers in the fourth detection layer enhances the overall capability of the model in capturing fine-grained feature descriptors. The detailed architecture of the modified model of YOLOV3 is depicted in Fig. 8. The four feature maps of this model have 13×13, 26×26, 52×52, and 104×104 sizes. Moreover, three additional anchor boxes were also introduced for this fourth detection layer. Overall, the 12 anchor boxes of this modified model of YOLOV3 are generated using the K means clustering technique. Because of the above additional features introduced into the modified model of YOLOV3, it has exceeded the performance of the original model for the detection of malarial parasites P. falciparum in microscopic thick blood smear images (Abdurahman et al., 2021).

Similarly, with a goal of YOLOV4 network architecture modification, one additional detection layer (fourth layer) was introduced to the existing detection layers, which is comprised of three detection layers, of the original model of YOLOV4 as shown in Fig. 7. The the new detection layer introduced layer enhances the performance of the modified model of YOLOV4 through the extraction of robust geometric and semantic feature descriptors by improving the feature extraction capability of the PANet model. For this model, a total of 12 anchor boxes with different sizes were generated and distributed evenly across the detection layers. The modified model of YOLV4, which has four detection layers, has been experimentally demonstrated to be more efficient in detecting Plasmodium falciparum than the original model of YOLOV4 (Abdurahman et al., 2021).

4 Challenges, Perspectives and Future Research Directions

The previous sections provided a review of techniques employed for automated malaria diagnosis in microscopic images of blood smears. We briefly described the conventional image processing and classical machine learning techniques, which were also covered by other existing studies (Umer et al., 2020). Our primary focus in this review was to survey the state-of-the-art (SOTA) deep learning-based approaches for automated malaria diagnosis. A comparison of the three approaches (conventional image processing, classical machine learning, and deep learning approaches) for malarial parasites' detection depicts that deep learning approaches outperform the other two approaches. The conventional image processing and classical machine learning approaches rely on manually designed feature extractors to detect objects of interest in images. The manually designed feature extractors and feature selection methods are suboptimal in many applications and require a high level of domain expertise in malarial parasites' detection. Hence, the generalization capabilities of these approaches are limited. They are also computationally intensive. Hence, the current perspective of automated malaria diagnosis system design is to use deep learning models.

Even though the SOTA deep learning models for object detection have achieved a good performance in malaria diagnosis applications, their implementation still needs further analysis. One of the challenges in developing these models is the lack of a large amount of data acquired in different laboratories and geographical regions. Because of this, pre-trained deep learning models are widely used. However, when we pre-train a model, the domain of the dataset for both the source and target should be somehow related. Unfortunately, the existing deep learning models for malaria diagnosis mostly use an ImageNet or COCO dataset as a source to pre-train the models. This is one of the limitations of the existing deep learning models. To improve the performance of deep learning models for malaria diagnosis, using pre-trained models trained using other histopathologic images would be one option that needs further investigation.

The other major issue pertinent to the existing architecture of the deep learning models for malaria diagnosis is that the focus of these pre-trained models was large object detections in non-medical application domains. Most of the existing SOTA deep learning models for object detection were designed to recognize large objects in natural images, such as those in ImageNet and COCO datasets. However, the size of malarial parasites in thick and thin blood microscopic images is very small. Thus, applying these architectures without considerable modification leads to low performance for detecting malarial parasites. It is difficult for the pre-trained models optimized to detect large objects to discriminate between staining artifacts in microscopic images and malarial parasites.

Another challenge that needs further investigation is that the existing deep learning models for malaria diagnosis are not easily explainable. The lack of explainability on how the models make decisions is frustrating for both model developers and clinicians. However, the lack of explainable deep learning models is not true only for malaria diagnosis-based studies, but it is also a common problem in many fields of computer vision in general. Hence, the design of explainable malarial parasites' detection models will be of much interest in the future.

To summarize, the future research direction on developing deep learning models for malaria diagnosis will be the following: (1) Refining the existing deep learning models to be more specific for small size object detection (malaria parasites). Image super-resolution algorithms can also be used as a preprocessing step to increase the size of malarial parasites in microscopic images. (2) Collection of large microscopic image datasets from different geographical locations to capture the variations in the blood smear staining conditions at different laboratory settings, (3) Design of explainable deep learning models for malarial parasites detection, (4) clinical validations and comprehensive comparative analysis of the existing malarial parasites' detection models.

5 Conclusion

In this chapter, a methodological review of machine learning techniques for malarial parasites detection and classification in stained blood film microscopic images was presented. Three types of deep learning methods were reviewed thoroughly. These are the patch-based CNN models, two-stage object detection models, and single-stage object detection models for detecting malarial parasites. The efficacy of these approaches in detection accuracy and computational complexity was thoroughly discussed. The single-stage object detection methods are the latest methods that are gaining popularity. These methods are also computationally efficient. However, the patch-based approach is the most computationally complex model. One of the challenges of the detection of malarial parasites in stained blood film microscopic images is that the parasites' size is small. And the original models of single-stage object detection techniques are not efficient for detecting small objects. With this regard, a modified algorithm was developed in (Abdurahman et al., 2021) to handle the small size object detection problems of the original models of YOLOV3 and YOLOV4. With the advancement of deep learning models for small size object detection in images, more accurate models are expected in the future. Future research on this topic also includes the reliability analysis of the different malarial parasites detection algorithms on various stained blood film microscopic images acquired at different laboratories in real clinical settings.

Acknowledgments

This research work was partially supported by a Transitional to Scale (TTS) grant titled "Mobile Phone Based Real Time Malaria and Tuberculosis Screening for Low Resource Settings in Ethiopia" from Armauer Hansen Research Institute (AHRI), Ministry of Health, The Federal Democratic Republic of Ethiopia.

References

Aayush Kumar, Sanat B. Singh, Suresh Chandra Satapathy and Minakhi Rout. (2021). Mosquito-net: A deep learning based cadx system for malaria diagnosis along with model interpretation using gradcam and class activation maps. *Expert Systems*, pp. e12695.

Abdul Nasir, A. S., Mashor, M. Y. and Mohamed, Z. (Dec 2012). Segmentation based approach for detection of malaria parasites using moving k-means clustering. In *2012 IEEE-EMBS Conference on Biomedical Engineering and Sciences*, pp. 653–658.

Aimon Rahman, Hasib Zunair, Mohammad Sohel Rahman, Jesia Quader Yuki, Sabyasachi Biswas et al. (2019). Improving malaria parasite detection from red blood cell using deep convolutional neural networks. *ArXiv, abs/1907.104189*.

Aimon Rahman, Hasib Zunair, Tamanna Rahman Reme, Sohel Rahman, M. and Mahdy, M. R. C. (2021). A comparative analysis of deep learning architectures on high variation malaria parasite classification dataset. *Tissue and Cell*, 69: 101473.

Alex Krizhevsky, Ilya Sutskever and Geoffrey E. Hinton. (2012). Imagenet classification with deep convolutional neural networks. In *NIPS*.

Alexey Bochkovskiy, Chien-Yao Wang and Hong-Yuan Mark Liao. (2020). Yolov4: Optimal speed and accuracy of object detection. *ArXiv, abs/2004.10934*.

Amin Siddiq Sumi, Hanung Adi Nugroho and Rudy Hartanto. (2019). A systematic review on automatic detection of plasmodium parasite. *International Journal of Engineering and Technology Innovation*, 11: 103–121.

András Vargha and Harold D. Delaney. (1998). The kruskal-wallis test and stochastic homogeneity. *Journal of Educational and Behavioral Statistics*, 23(2): 170–192.

Andrea Loddo, Cecilia Di Ruberto and Michel Kocher. (2018). Recent advances of malaria parasites detection systems based on mathematical morphology. *Sensors (Basel, Switzerland)*, 18, 02.

Andrew L. Maas. (2013). Rectifier nonlinearities improve neural network acoustic models.

Boray Tek, F., Andrew Graham Dempster and Izzet Kale. (2010). Parasite detection and identification for automated thin blood film malaria diagnosis. *Computer Vision and Image Understanding*, 114(1): 21–32.

Chaobo Zhang, Chih-Chen Chang and Maziar Jamshidi. (2018). Bridge damage detection using single-stage detector and field inspection images. *CoRR, abs/1812.10590*.

Charles B. Delahunt, Mayoore S. Jaiswal, Matthew P. Horning, Samantha Janko, Clay M. Thompson et al. (2019). Fully-automated patient-level malaria assessment on field-prepared thin blood film microscopy images, including supplementary information. *CoRR, abs/1908.01901*.

Courosh Mehanian, Mayoore Jaiswal, Charles Delahunt, Clay Thompson, Matt Horning et al. (2017). Computer-automated malaria diagnosis and quantitation using convolutional neural networks. In *Proceedings of the IEEE International Conference on Computer Vision Workshops*, pp. 116–125.

David Mas, Belen Ferrer, Dan Cojoc, Sara Finaurini, Vicente Mico et al. (2015). Novel image processing approach to detect malaria. *Optics Communications*, 350: 13–18.

David Pan, W., Yuhang Dong and Dongsheng Wu. (2018). Classification of malaria-infected cells using deep convolutional neural networks. *In*: Hamed Farhadi (ed.). *Machine Learning*, Chapter 8. IntechOpen, Rijeka.

De Rong Loh, Wen Xin Yong, Jullian Yapeter, Karupppasamy Subburaj and Rajesh Chandramohanadas. (2021). A deep learning approach to the screening of malaria infection: Automated and rapid cell counting, object detection and instance segmentation using mask r-cnn. *Computerized Medical Imaging and Graphics*, 88: 101845.

Dev Kumar Das, Madhumala Ghosh, Mallika Pal, Asok K. Maiti and Chandan Chakraborty. (2013). Machine learning approach for automated screening of malaria parasite using light microscopic images. *Micron*, 45: 97–106.

Dhanya Bibin, Madhu S. Nair and Punitha, P. (2017). Malaria parasite detection from peripheral blood smear images using deep belief networks. *IEEE Access*, 5: 9099–9108.

Dong, Y., Jiang, Z., Shen, H., David Pan, W., Williams, L. A. et al. (Feb 2017). Evaluations of deep convolutional neural networks for automatic identification of malaria infected cells. In *2017 IEEE EMBS International Conference on Biomedical Health Informatics (BHI)*, pp. 101–104.

Eric Hanssen, Kenneth N. Goldie and Leann Tilley. (2010). Ultrastructure of the asexual blood stages of plasmodium falciparum. *Methods in Cell Biology*, 96: 93–116.

Fan chiang Yang, Nicolas Quizon, Kamolrat Silamut, Richard James Maude, Stefan Jäger et al. (2019). Cascading yolo: Automated malaria parasite detection for plasmodium vivax in thin blood smears.

Feng Yang, Mahdieh Poostchi, Hang Yu, Zhou Zhou, Kamolrat Silamut et al. (2019). Deep learning for smartphone-based malaria parasite detection in thick blood smears. *IEEE Journal of Biomedical and Health Informatics*, pp. 1–1, 09.

Fetulhak Abdurahman, Kinde Anlay Fante and Mohammed Aliy. (2021). Malaria parasite detection in thick blood smear microscopic images using modified yolov3 and yolov4 models. *BMC Bioinformatics*, 22(1): 1–17.

Girshick, R. (2015). Fast r-cnn. In *2015 IEEE International Conference on Computer Vision (ICCV)*, pp. 1440–1448.

Girshick, R., Donahue, J., Darrell, T. and Malik, J. (2014). Rich feature hierarchies for accurate object detection and semantic segmentation. In *2014 IEEE Conference on Computer Vision and Pattern Recognition*, pp. 580–587.

Gopalakrishna Pillai Gopakumar, Murali Swetha, Gorthi Sai Siva and Gorthi R. K. Sai Subrahmanyam. (2018). Convolutional neural network-based malaria diagnosis from focus stack of blood smear images acquired using custom-built slide scanner. *Journal of Biophotonics*, 11: 3.

Han Sang Park, Matthew T. Rinehart, Katelyn A. Walzer, Jen-Tsan Ashley Chi and Adam Wax. (2016). Automated detection of P. falciparum using machine learning algorithms with quantitative phase images of unstained cells. *PLOS ONE*, 11(9): 1–19, 09.

Hang Yu, Feng Yang, Sivaramakrishnan Rajaraman, Ilker Ersoy, Golnaz Moallem et al. (2020). Malaria screener: A smartphone application for automated malaria screening. *BMC Infectious Diseases*, 20: 11.

Jane Hung, Allen Goodman, Stefanie Lopes, Gabriel Rangel, Deepali Ravel et al. (2017). Applying faster r-cnn for object detection on malaria images. *ArXiv,abs/1804.09548*.

Jasper Uijlings, Sande, K., Gevers, T. and Arnold Smeulders. (2013). Selective search for object recognition. *International Journal of Computer Vision*, 104: 154–171, 09.

John Quinn, Alfred Andama, Ian Munabi and Fred Kiwanuka. (2014). Automated blood smear analysis for mobile malaria diagnosis, pp. 115–132, 09.

Joseph Redmon and Ali Farhadi. (2016). YOLO9000: better, faster, stronger. *CoRR, abs/1612.08242*.

Joseph Redmon and Ali Farhadi. (2018). Yolov3: An incremental improvement. *CoRR, abs/1804.02767*.

Kaiming He, Xiangyu Zhang, Shaoqing Ren and Jian Sun. (2014). Spatial pyramid pooling in deep convolutional networks for visual recognition. pp. 346–361. *In*: David Fleet, Tomas Pajdla, Bernt Schiele and Tinne Tuytelaars (eds.). *Computer Vision—ECCV 2014*, Cham, 2014. Springer International Publishing.

Karen Simonyan and Andrew Zisserman. (2015). Very deep convolutional networks for large-scale image recognition. *CoRR, abs/1409.1556*.

Katherine Torres, Christine M. Bachman, Charles B. Delahunt, Jhonatan Alarcon Baldeon, Freddy Alava et al. (Sep 2018). Automated microscopy for routine malaria diagnosis: A field comparison on giemsa-stained blood films in peru. *Malaria Journal*, 17(1): 339.

Kristofer Delas Peñas, Pilarita Rivera and Prospero Naval. (2018). *Analysis of Convolutional Neural Networks and Shape Features for Detection and Identification of Malaria Parasites on Thin Blood Smears*, pp. 472–481.

Li Liu, Wanli Ouyang, Xiaogang Wang, Paul W. Fieguth, Jie Chen et al. (2018). Deep learning for generic object detection: A survey. *CoRR, abs/1809.02165*.

Luís Rosado, José M. Correia da Costa, Dirk Elias and Jaime S. Cardoso. (2016). Automated detection of malaria parasites on thick blood smears via mobile devices. *Procedia Computer Science*, 90: 138–144. 20th Conference on Medical Image Understanding and Analysis (MIUA 2016).

Mahdieh Poostchi, Ilker Ersoy, Katie McMenamin, Emile Gordon, Nila Palaniappan et al. (2018). Malaria parasite detection and cell counting for human and mouse using thin blood smear microscopy. *Journal of Medical Imaging*, 5(4): 1–13.

Mahdieh Poostchi, Kamolrat Silamut, Richard Maude, Stefan Jaeger and George Thoma. (2018). Image analysis and machine learning for detecting malaria. *Translational Research*, 194, 01.

Mingxing Tan, Ruoming Pang and Quoc V. Le. (2019). Efficientdet: Scalable and efficient object detection. *ArXiv,abs/1911.09070*.

Muhammad Umer, Saima Sadiq, Muhammad Ahmad, Saleem Ullah, Gyu Sang Choi et al. (2020). A novel stacked cnn for malarial parasite detection in thin blood smear images. *IEEE Access*, 8: 93782–93792.

Nicholas E. Ross, Charles J. Pritchard, David M. Rubin and Adriano G. Dusé. (May 2006). Automated image processing method for the diagnosis and classification of malaria on thin blood smears. *Medical and Biological Engineering and Computing*, 44(5): 427–436.

Olga Russakovsky, Jia Deng, Hao Su, Jonathan Krause, Sanjeev Satheesh et al. (2015). Imagenet large scale visual recognition challenge. *International Journal of Computer Vision*, 115: 211–252.

Olusola Ojurongbe, Olunike Olayeni Adegbosin, Sunday Samuel Taiwo, Oyebode Armstrong Terry Alli, Olugbenga Adekunle Olowe et al. (2013). Assessment of clinical diagnosis, microscopy, rapid diagnostic tests, and polymerase chain reaction in the diagnosis of plasmodium falciparum in nigeria. *Malaria Research and Treatment*, 2013.

Poonguzhali Elangovan and Malaya Nath. (2021). A novel shallow convnet-18 for malaria parasite detection in thin blood smear images: Cnn based malaria parasite detection. *SN Computer Science*, 2: 09.

Rajaraman Sivaramakrishnan, Sameer Antani and Stefan Jaeger. (Aug 2017). Visualizing deep learning activations for improved malaria cell classification. pp. 40–47. *In*: Samah Fodeh and Daniela Stan Raicu (eds.). *Proceedings of the First Workshop Medical Informatics and Healthcare held with the 23rd SIGKDD Conference on Knowledge Discovery and Data Mining*, volume 69 of Proceedings of Machine Learning Research. PMLR, 14.

Redmon, J., Divvala, S., Girshick, R. and Farhadi, A. (2016). You only look once: Unified, real-time object detection. In *2016 IEEE Conference on Computer Vision and Pattern Recognition (CVPR)*, pp. 779–788.

Ren, S., He, K., Girshick, R. and Sun, J. (2017). Faster r-cnn: Towards real-time object detection with region proposal networks. *IEEE Transactions on Pattern Analysis and Machine Intelligence*, 39(6): 1137–1149.

Richard O. Duda and Peter E. Hart. (1972). Use of the hough transformation to detect lines and curves in pictures. *Commun. ACM*, 15: 11–15.

Rose Nakasi, Ernest Mwebaze, Aminah Zawedde, Jeremy Tusubira, Benjamin Akera et al. (2020). A new approach for microscopic diagnosis of malaria parasites in thick blood smears using pre-trained deep learning models. *SN Applied Sciences*, 2(7): 1–7.

Salam Shuleenda Devi, Ngangbam Herojit Singh and Rabul Hussain Laskar. (2019). Performance analysis of various feature sets for malaria-infected erythrocyte detection. pp. 275–283. *In*: Kedar Nath Das, Jagdish Chand Bansal, Kusum Deep, Atulya K. Nagar, Ponnambalam Pathipooranam et al. (eds.). *Soft Computing for Problem Solving*. Singapore, 2019. Springer Singapore.

Salam Shuleenda Devi, Rabul Hussain Laskar and Shah Alam Sheikh. (Apr 2018). Hybrid classifier based life cycle stages analysis for malaria-infected erythrocyte using thin blood smear images. *Neural Computing and Applications*, 29(8): 217–235.

Samson Chibuta and Aybar Can Acar. (2020). Real-time malaria parasite screening in thick blood smears for low-resource setting. *Journal of Digital Imaging*, pp. 1–13.

Sanghamitra Sathpathi, Akshaya K. Mohanty, Parthasarathi Satpathi, Saroj K. Mishra, Prativa K. Behera et al. (2014). Comparing leishman and giemsa staining for the assessment of peripheral blood smear preparations in a malaria-endemic region in India. *Malaria Journal*, 13(1): 1–5.

Sergey Ioffe and Christian Szegedy. (2015). Batch normalization: Accelerating deep network training by reducing internal covariate shift.

Seunghyun Moon, Sukjun Lee, Heechang Kim, Lucio H. Freitas-Junior, Myungjoo Kang et al. (2013). An image analysis algorithm for malaria parasite stage classification and viability quantification. *PLOS ONE*, 8(4): 1–12, 04.

Shu Liu, Lu Qi, Haifang Qin, Jianping Shi and Jiaya Jia. (2018). Path aggregation network for instance segmentation. *CoRR, abs/1803.01534*.

Sivaramakrishnan Rajaraman, Sameer K. Antani, Mahdieh Poostchi, Kamolrat Silamut, Md A. Hossain et al. (2018). Pre-trained convolutional neural networks as feature extractors toward improved malaria parasite detection in thin blood smear images. *PeerJ*, 6: e4568.

Tek, F., Andrew Dempster and Izzet Kale. (2009). Computer vision for microscopy diagnosis of malaria. *Malaria Journal*, 8: 153, 02.

Tsung-Yi Lin, Piotr Dollár, Ross B. Girshick, Kaiming He, Bharath Hariharan et al. (2016). Feature pyramid networks for object detection. *CoRR, abs/1612.03144*.

Ugur N. Kucuk, Mehmet Eyuboglu, Hilal Olgun Kucuk and Gokhan Degirmencioglu. (2016). Importance of using proper post hoc test with anova. *International Journal of Cardiology*, 209: 346.

Vijayalakshmi and Rajesh Kanna. (Jan 2019). Deep learning approach to detect malaria from microscopic images. *Multimedia Tools and Applications*.

Wei Liu, Dragomir Anguelov, Dumitru Erhan, Christian Szegedy, Scott Reed et al. (2016). Ssd: Single shot multibox detector. pp. 21–37. *In*: Bastian Leibe, Jiri Matas, Nicu Sebe and Max Welling (eds.). *Computer Vision—ECCV 2016*, Cham, 2016. Springer International Publishing.

Wendy O'Meara, Mazie Barcus, Chansuda Wongsrichanalai, Muth Sinuon, Jason Maguire et al. (2006). Reader technique as a source of variability in determining malaria parasite density by microscopy. *Malaria Journal*, 5: 118, 02.

World Health Organization et al. (2020). World malaria report 2019. 2019. Reference Source. https://www.who.int/malaria/publications/world-malaria-report-2019/en.

Yang, F., Poostchi, M., Yu, H., Zhou, Z., Silamut, K. et al. (2019). Deep learning for smartphone-based malaria parasite detection in thick blood smears. *IEEE Journal of Biomedical and Health Informatics*, pp. 1–1.

Yann LeCun, Léon Bottou, Yoshua Bengio and Patrick Haffner. (1998). Gradient-based learning applied to document recognition.

Yashasvi Purwar, Sirish Shah, Gwen Clarke, Areej Almugairi and Atis Muehlenbachs. (2011). Automated and unsupervised detection of malaria parasites in microscopic images. *Malaria Journal*, 10: 364, 12.

Yasmin Kassim, Palaniappan, K., Feng Yang, Mahdieh Poostchi, Nila Palaniappan et al. (2020). Clustering-based dual deep learning architecture for detecting red blood cells in malaria diagnostic smears. *IEEE Journal of Biomedical and Health Informatics*, pp. 1–1, 10.

Zhaohui Liang, Andrew Powell, Ilker Ersoy, Mahdieh Poostchi, Kamolrat Silamut et al. (2016). Cnn-based image analysis for malaria diagnosis. *2016 IEEE International Conference on Bioinformatics and Biomedicine (BIBM)*, pp. 493–496.

Zhaohui Liang, Andrew Powell, Ilker Ersoy, Mahdieh Poostchi, Kamolrat Silamut et al. (2016). Cnn-based image analysis for malaria diagnosis. In *2016 IEEE International Conference on Bioinformatics and Biomedicine (BIBM)*, pp. 493–496. IEEE.

Chapter 11

Automatic Classification of Coronary Stenosis using Convolutional Neural Networks and Simulated Annealing

Luis Diego Rendon-Aguilar,[1] *Ivan Cruz-Aceves,*[2,]* *Arturo Alfonso Fernandez-Jaramillo,*[1] *Ernesto Moya-Albor,*[3,]* *Jorge Brieva*[3] and *Hiram Ponce*[3]

ABSTRACT

Automatic detection of coronary stenosis plays an essential role in systems that perform computer-aided diagnosis in cardiology. Coronary stenosis is a narrowing of the coronary arteries caused by plaque that reduces the blood flow to the heart. Automatic classification of coronary stenosis images has been re-

---

[1] Universidad Politécnica de Sinaloa, Carretera Municipal Libre Mazatlán Higueras km 3, Col. Genaro Estrada, Mazatlán, Sinaloa, México.
[2] CONACYT - Centro de Investigación en Matemáticas (CIMAT), A.C. Jalisco S/N, Col. Valenciana, C.P. 36000, Guanajuato, Gto, México.
[3] Universidad Panamericana, Facultad de Ingeniería, Augusto Rodin 498, Ciudad de México, 03920, México.
* Corresponding authors: cruz@cimat.mx

cently addressed using deep and machine learning techniques. Generally, the machine learning methods form a bank of empirical and automatic features from the angiographic images. In the present work, a novel method for the automatic classification of coronary stenosis X-ray images is presented. The method is based on convolutional neural networks, where the neural architecture search is performed by using the path-based metaheuristics of simulated annealing. To perform the neural architecture search, the maximization of the F1-score metric is used as the fitness function. The automatically generated convolutional neural network was compared with three deep learning methods in terms of the accuracy and F1-score metrics using a testing set of images obtaining 0.88 and 0.89, respectively. In addition, the proposed method was evaluated with different sets of coronary stenosis images obtained via data augmentation. The results involving a number of different instances have shown that the proposed architecture is robust preserving the efficiency with different datasets.

1 Introduction

The automatic detection of coronary stenosis plays an essential role in systems that perform computer-aided diagnosis in cardiology. The coronary stenosis is a narrowing of the coronary arteries caused by plaque that reduces the blood flow to the heart. In general, the process uses X-ray coronary angiograms, which has two main disadvantages: non-uniform illumination and low contrast between vessels and background images. Another drawback in terms of computational implementation, is the small number of available public datasets of coronary stenosis, and the low rate of true positive cases obtained with respect to false positive cases that are acquired from each X-ray image or video sequence.

The automatic classification of coronary stenosis images has been recently addressed using deep and machine learning techniques. Generally, the methods based on machine learning form a bank of empirical and automatic features from the angiographic images. The automatic features are extracted using vessel enhancement or vessel segmentation methods, which are previously tuned. Moreover, the process of feature extraction is avoided through the use of deep learning methods. Nevertheless, there are some issues to keep in mind, for example, computational time and computational resources, the data augmentation techniques used to increase the number of samples, lack of knowledge about the main features for solving the problem, and a weak exploration of the search space for finding an optimal neural architecture.

Recent works have used machine learning for stenosis classification. Cruz-Aceves et al. (Cruz-Aceves et al., 2017) used a Naive Bayes method to detect stenosis cases based on the features like the sum of vessel widths, mean of the vessel width histogram, and standard deviation of the vessel width histogram. Gil-Rios et al. (Gil-Rios et al., 2021) defined a bank of 49 computational features to be selected using a feature selection strategy based on the estimation of distribution algorithms and support vector machines. Moreover, Antczak and Liberadzki (Antczak and Liberadzki, 2018) introduced deep neural networks for stenosis classification, where a data augmentation strategy based on splines as synthetic vessels was implemented. Ovalle-Magallanes et al., 2020 proposed a Convolutional Neural Network (CNN) based on the strategy of transfer learning. In this method, the pre-trained VGG16, ResNet50, and Inception-v3 neural networks were adopted, and a fine tuning step was implemented for the binary classification of coronary stenosis. Cong et al. (Cong et al., 2019) proposed a deep-learning based workflow for stenosis classification with a categorization of three different groups. Danilov et al., 2021 introduced a comparative study with eight different neural network architectures using clinical angiography data of 100 patients and the metrics of F1-score and mean average precision. Moon et al. (Moon et al., 2021) proposed the use of convolutional neural networks based on a self-attention mechanism. which is evaluated in terms of the average area under the curve.

In general, the neural architecture of the aforementioned methods based on convolutional neural networks has been empirically determined. To solve the problem of automatic neural architecture search, different Evolutionary Computation (EC) methods have been introduced (Liu et al., 2021). The most explored EC methods are genetic algorithms (Sun et al., 2020; Xie and Yuille, 2017; Lu, 2020), genetic programming (Suganuma et al., 2018), harmony search (Kim et al., 2020), global optimization (Stein et al., 2019), and particle swarm optimization (Passricha et al., 2019; Wang et al., 2020; Fernandes Junior and Yen, 2019; Yamasaki et al., 2017; Serizawa and Fujita, 2020; Wang et al., 2018; Sun et al., 2019; Qolomany et al., 2017). Since the EC methods are population-based on a set of potential solutions or individuals (CNN architectures, also called proxies CNN), the evaluation of the fitness value for each individual is an expensive and time-consuming task. To address this computational drawback, path-based metaheuristics can explore the high-dimensional search space for the neural architecture search of CNNs.

In this chapter, a novel method for the automatic classification of coronary stenosis X-ray images is presented. The method is based on convolutional neural networks, where the architecture neural search is performed by using the path-based metaheuristic of Simulated Annealing (SA). In the training stage, the optimization process is carried out using a training set of coronary stenosis image patches, and in the testing step, the automatically defined neural architecture is applied to a testing set of images. To perform the neural architecture search, the maximization of the F1-score metric is used as the fitness function.

Finally, the performance of the proposed method is compared using different data subsets using a data augmentation strategy.

The organization of the chapter is as follows. Section 2 introduces the fundamentals of the convolutional neural networks for image classification and the metaheuristic of simulated annealing for combinatorial optimization. The proposed method based on SA for the automatic neural architecture search is presented in Section 3. The computational experiments of the proposed method and its comparison with different state of the art methods are presented in Section 4, and finally, conclusions are given in Section 5.

2 Background

In the present section, the fundamentals of convolutional neural networks and the metaheuristic of Simulated annealing are explained in detail. These two methods are of particular interest because they are used in the proposed method. In addition, since SA is used for the optimization of the neural architecture search, the Traveling Salesman Problem (TSP) is used as a practical application.

2.1 Convolutional Neural Networks

The convolutional neural network is a recent technology that has obtained efficient results in different areas such as object recognition or computer vision. In addition, it has been used in fields such as medicine, biology, and the entertainment industry. To understand how CNN works, the concept of neurons and fully connected layers is illustrated in Fig. 1.

A single neuron represents a transformation function, in which a set of data is introduced. In general, a neuron can be defined as follows:

$$\sum_{i=0}^{n}(w_i + x_i) + B, \qquad (1)$$

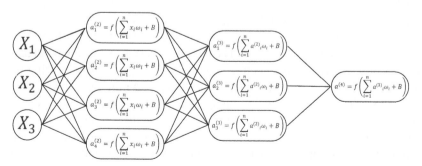

Figure 1: Basic model of a fully connected neural network.

where w is a weight value, x is the input data, and B is known as the bias value. In the optimization process, the values w and B must be tuned by using a predefined method on all the neurons that belong to a layer. In this process, each neuron can be seen as a linear regression problem, in which the number of epochs, or convergence rate can be used as a stopping criterion. In Fig. 2, the general representation of a single neuron is presented.

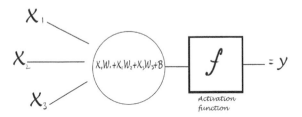

Figure 2: Representation of a single artificial neuron.

In a neuron, the activation function represents the function for mapping data. In general, the most commonly used activation functions are the sigmoid and ReLU (Rectified Linear Unit). Figure 3 presents the behavior of these activation functions.

The classical artificial neural networks (ANNs) have been an important machine learning algorithm because of their predicting efficiency and finding hidden patterns in low dimensional data. For working problems involving high-dimensional data, convolutional neural networks have proven to be an efficient deep learning technique. CNNs use the mathematical convolution operator to extract different features from the input image by applying a number

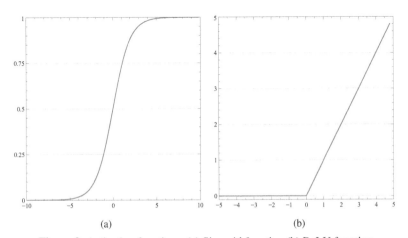

Figure 3: Activation functions: (a) Sigmoid function (b) ReLU function.

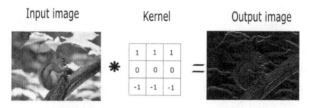

Figure 4: Result of applying a convolution kernel on an input image obtaining an edge detector.

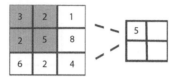

Figure 5: Process of the max pooling layer for reducing a feature map.

of kernels or filters. In Fig. 4, the convolution operation for extracting edge pixels is shown.

To define the filters for convolution, parameters such as size and number of filters and stride (how the filter moves around the image) have to be tuned. Another CNN mathematical operator is called max pooling, which is used to reduce the output size from the previous convolution, while saving computational time. In Fig. 5, the process of the max pooling operator is shown, where in from a specific area of pixels, the pixel with the highest value is preserved to form the new matrix.

The CNNs are represented by sequential layers, taking the output of the previous one for computing new features. In general, although the number of layers needs to be empirically predefined to solve particular problems, it is well known that initial layers extract basic information from the data, and deeper layers extract more complex features. In Fig. 6, the generic representation of a CNN involving feature learning and the classification steps is presented.

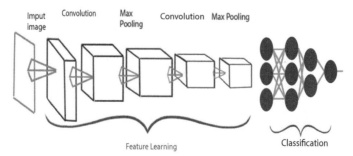

Figure 6: Generic architecture of a convolutional neural network.

2.2 Simulated Annealing

Simulated annealing is a single-solution stochastic strategy used for global optimization introduced by Kirkpatrick et al., 1983. SA is based on the idea of controlling the cooling of a material after heating it to its melting point by altering its physical properties. SA explores during the iterative process different potential solutions to the combinatorial problem and it accepts a new solution adopting an elite strategy, however, a difference with respect to brute force methods, SA can accept *worse* solutions to avoid local minima (Talbi, 2009). The process to accept new solutions directly depends on the current temperature parameter and the ΔE parameter which is obtained from the objective function. These two parameters are used to compute the probability of acceptance based on the Boltzmann distribution as follows:

$$P(\Delta E, T) = \exp\left(-\left(\frac{f(s') - f(s)}{T}\right)\right), \qquad (2)$$

where ΔE represents the difference between the current and the new potential solution and T represents temperature. The probability of accepting solutions that do not improve the fitness of the optimization search is governed by the parameter T, which is gradually decreased in such a way that only improved solutions are accepted when the temperature is descending (Dowsland and Thompson, 2012). This process is iteratively performed until the stopping or equilibrium criterion is satisfied. A comparative instance of how an exhaustive search method works and how the main idea behind simulated annealing is performed is illustrated in Fig. 7. On the other hand, since the simulated annealing is mainly governed by the temperature parameter, in Fig. 8, the behavior of this parameter is introduced along with the probability of acceptance. The data sets for these figures were obtained from the traveling salesman problem, which is described in the following Section 2.2.1. In Algorithm 1, an implementation of the simulated annealing method is shown.

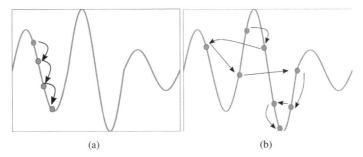

(a)   (b)

Figure 7: (a) Exhaustive search strategy. (b) Simulated Annealing strategy.

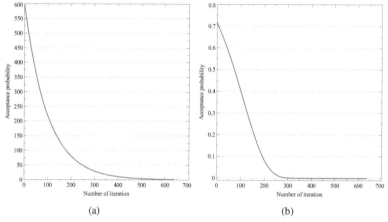

Figure 8: Iterative process of SA: (a) Temperature parameter, and (b) Acceptance probability.

Algorithm 1 Simulated Annealing

Require: $D =$ Number of dimensions, $Temp =$ Temperature
1: Initialize $t = 0$, $X^t \sim U(0,1)$
2: **while** stoppingCriterion $\neq$ **true do**
3: $\quad X_*^t = Perturb(X^t)$
4: $\quad X^t = Accept(X^t, X_*^t)$, according to $\exp^{-\left(\frac{\Delta}{Temp}\right)}$
5: $\quad Temp = Temp * \alpha$
6: **end while**
7: **return** $X^t, f(X^t)$

2.2.1 Application: Traveling Salesman Problem

The traveling salesman problem is a combinatorial optimization problem consisting of the idea of a travelling salesman that has to visit different cities, in the most efficient way, using the minimal distance.

The TSP represents a non-deterministic polynomial time problem (NP-Problem), since the high number of permutations can not be tested with conventional computing resources. For instance, a TSP with only 10 cities has 3,628,800 possible solutions. To solve the TSP, a number of different strategies have been introduced such as genetic algorithms (Razali, 2015), approximation algorithms (Williamson and Shmoys, 2010), and simulated annealing (Song et al., 2003).

Since the metaheuristic of SA can be used in different problems preserving the main idea of the method, here the TSP problem is used to illustrate the performance of SA in combinatorial optimization problems. The source code of the simulated annealing method for solving the TSP problem is provided in the Appendix 5, using the Python 3 programming language.

The TSP solved in the present section, consists of 50 randomly generated cities in the range $[0, 200]$. In Fig. 9, the problem is illustrated, where each city is representing by its x and y coordinates, and a random solution is also shown.

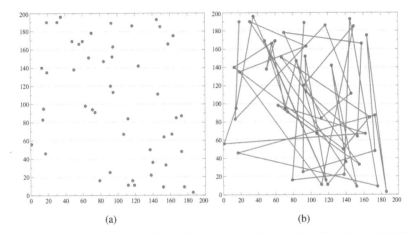

Figure 9: TSP problem: (a) Position in Cartesian coordinates of the 50 cities, and (b) Random route as possible solution.

In the TSP problem, the most common fitness or cost function is the total distance traveled by the salesman, which is the value that has to be minimized as much as possible. In general the total distance is computed used the Euclidean distance as follows:

$$d(p,q) = \sqrt{\sum_{i=1}^{n} (q_i - p_i)^2}, \qquad (3)$$

where n, is the number of cities of the problem, and p and q represent two points or cities in the Euclidean n-space.

Moreover, the selection of the initial temperature in the SA method, is not a trivial task. Different strategies have been proposed in a general context by Ben-Ameur (Ben-Ameur, 2004), and specific strategies for measuring the variance in cost for small samples have been proposed by Lewis (Lewis, 2007). The temperature and cooling rate can be viewed as a number of iterations in another context. Consequently, a proper tuning of these parameters is relevant to the problem, since a low number of iterations can cause a poor exploration of the high-dimensional space, and a high number of iterations drastically increases the computational time of the optimization process. The cooling rate is a value in the range $(0, 0.99]$ that defines the rate for decreasing the temperature, when the cooling rate is low the exploration is high, reducing the ability of finding optimal solutions.

The parameter values used in the experiments were set as $Cooling = 0.99$ and the $Temperature$ value was determined according to (Lewis, 2007), using the standard deviation of the fitness of 25 random solutions. Since the SA is a stochastic optimization method, in Table 1, the fitness of 5 independent trials is presented. Finally, in Fig. 10, the best solution (best route) obtained by simulated annealing is illustrated, along with the searching process through the high-dimensional space performed during the iterative strategy.

Table 1: Independent experiments of SA for solving the TSP with 50 cities.

Run	Best route (Cost)
1	1125.98
2	**1121.92**
3	1193.82
4	1160.17
5	1214.49

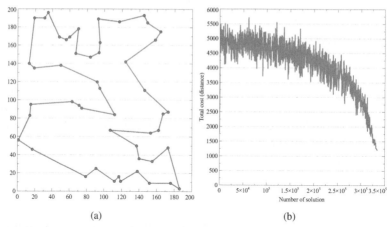

Figure 10: Solution of the TSP obtained by SA: (a) Route with the lowest distance, and (b) Searching process performed by SA.

3 Proposed Method

In the present section, the proposed method for automatic neural architecture search is presented along with the metrics for evaluating the performance of the automatically generated convolutional neural network.

The automatic neural architecture search in convolutional neural networks is a challenging problem because of the computational complexity. In the problem, the parameters of a number of convolutional layers, number and size of filters, number of max pooling layers, dropout value, learning rate or number of epochs, just to mention a few, have to be determined. This high-dimensional search space is explored by simulated annealing to find a set of parameters that represents a global optimum. Generally, in high-dimensional problems, the optimum is not obtained, but using metaheuristics an efficient solution can be acquired.

In the proposed method, a proxy CNN (CNN with a potential architecture) is generated by using an initial random configuration. During the optimization process performed by SA, in each evaluation a proxy CNN is generated and evaluated by using a fitness function. The fitness of each proxy CNN is computed using the F1-score metric, which has to be maximized. Along the SA optimization process, several proxy CNNs are randomly generated and the proxy CNN with the best fitness is preserved as the optimal architecture.

In literature, different metaheuristics for global optimization can be applied to the present problem. The main advantage of using a path-based metaheuristic instead of a population-based method is the computational time. Since, for each evaluation of the fitness function, a proxy CNN has to be generated from scratch, the computational time is highly expensive. Consequently, a path-based metaheuristic such as SA, which only uses one proxy CNN at a time, reduces the computational time in obtaining an efficient architecture.

In order to evaluate the performance of the proposed method, the accuracy and F1-score metrics have been adopted. The accuracy metric is computed as follows:

$$Accuracy = \frac{TP+TN}{TP+FP+TN+FN}, \qquad (4)$$

where TP is the true-positive rate consisting of the number of correctly classified stenosis cases, TN is the true-negative rate, which is the number of correctly classified stenosis cases, and FP and FN are the false-positive and false-negative rates involving the number of incorrectly classified stenosis and no stenosis cases, respectively. Moreover, the F1-score metric, which is also used to measure the performance of a classification method is computed as follows:

$$F1 = \frac{2TP}{2TP+FP+FN}. \qquad (5)$$

Both evaluation metrics have to be maximized and they are in the range $[0,1]$, where 1 is the optimal classification and 0 is the worst case. These metrics are used for the evaluation of the proposed CNN with automatic neural architecture search and comparative methods in terms of classification performance. In addition, the F1-score has also been used as the fitness function to evaluate the potential solutions (proxy CNNs) generated via simulated annealing and the training set of coronary stenosis images. The proxy CNN with the highest F1-score in the training step, represents the proposed CNN that is directly applied on the test set of X-ray stenosis images.

4 Computational Experiments

In this section, the computational experiments of the proposed method are presented and analyzed using a public database of coronary stenosis images. The dataset of coronary stenosis used in the present research was introduced by Antczak and Liberadzki, 2018 consisting of 125 real positive cases and 125 selected negative cases of size 32×32 pixels. For different experiments, data augmentation was applied, and the set of images was divided into the training and testing sets using 33.3% and 66.6%, respectively. In the training set, the data was divided into training and validation subsets using a rate of 90% − 10%. Moreover, the experiments were performed using the programming language Python 3, and a PC graphics card NVidia Titan RTX with 24GB of RAM.

In order to illustrate the coronary stenosis X-ray images used in the experiments, a subset of true positive and negative patches are presented in Fig. 11.

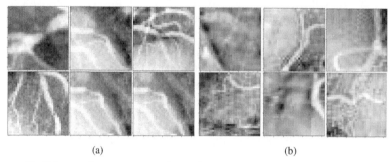

(a)   (b)

Figure 11: Subsets of coronary stenosis patches: (a) True positive instances, and (b) True negative instances.

Since the proposed method is based on a convolutional neural network automatically defined using simulated annealing a proper parameter tuning of SA is required. Different experiments were performed using a range of values for the temperature and cooling rate parameters. In Table 2, the set of parameters with the best performance in the training step are introduced.

In Table 3, the automatically obtained CNN architecture is presented. The CNN uses a batch size of 64, a learning rate of 0.01, a *momentum* = 0.8 and 800 number of epochs. This CNN was trained specifically for the coronary

Table 2: Set of Simulated annealing parameters used in the proposed method.

Parameter	Value
Temperature	3.0
Cooling rate	0.95
Iterations per temperature	15

Table 3: Best CNN architecture obtained using simulated annealing.

Layer	Filter size	Stride	Drop out
16 CONV2D	3×3	1×1	0.5
MAX POOL	2×2	1×1	–
16 CONV2D	5×5	1×1	–
MAX POOL	2×2	1×1	–
8 CONV2D	5×5	1×1	0.5
16 CONV2D	1×1	2×2	0.5
8 FC/RELU	–	–	–
1 FC/SIGMOID (OUTPUT)	–	–	–

Table 4: Comparative analysis of the proposed method with different methods of the state of the art using the testing set.

Method	Accuracy	F1-Score
MobileNetv2	0.60	0.62
Antczak (Antczak and Liberadzki, 2018)	0.73	0.67
LeNet5	0.74	0.78
Proposed method	**0.88**	**0.89**

stenosis problem using the aforementioned database, avoiding a transfer learning strategy. Comparative analysis with different methods of the state of the art is presented in Table 4. This Table shows that the proposed method outperforms the comparative methods in terms of the accuracy and F1-score metrics. Additionally, the comparative methods were trained using the training set of coronary stenosis cases and preserving the parameter values recommended for authors.

On the other hand, additional experiments were carried out on the proposed method using a data augmentation strategy. Sets of different number of stenosis instances were artificially generated and evaluated in terms of the F1-score. In Fig. 12, the performance of the proposed method is presented. In this experiment, the classification results show that the automatically generated CNN architecture is robust for application with a different number of instances, which is desirable in most of the classification problems. Finally, in Fig. 13, subsets of images correctly and incorrectly classified as stenosis/no stenosis by the proposed method are presented. In deep learning techniques it is a common practice to avoid the use of preprocessing steps, in these instances, the false positive and false negative images are in general isolated vessels with a nonuniform illumination. This drawback can be improved in future works by introducing a Hessian or Gaussian-based enhancement method prior to the classification task.

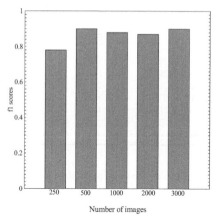

Figure 12: Results on different sets of coronary stenosis using data augmentation.

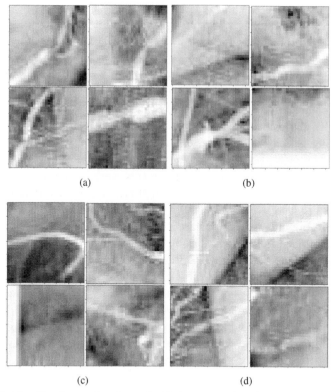

Figure 13: Classification results obtained with the proposed method: (a) True positives, (b) True Negatives, (c) False positives, and (d) False negatives.

5 Conclusions

In the present chapter, a novel method for the automatic classification of coronary stenosis in X-ray angiograms based on convolutional neural networks and the metaheuristic simulated annealing has been introduced. The proposed method uses SA in the training step for automatic neural architecture search. In this step, the dataset of coronary stenosis images was divided into training and validation subsets preserving a 90%–10% rate. The automatically generated convolutional neural network was compared with three deep learning methods in terms of the accuracy and F1-score metrics using a testing set of images obtaining 0.88 and 0.89, respectively. In addition to the experimental results, the proposed CNN was evaluated with different sets of coronary stenosis images obtained via data augmentation. The results with sets involving different number of instances have shown that the proposed CNN is robust in preserving the efficiency with different datasets. This advantage of the proposed method is relevant for consideration in systems that perform computer-aided diagnosis in cardiology.

Acknowledgments

This research has been supported by the National Council of Science and Technology of México under the project: Cátedras-CONACYT 3150-3097.

Ernesto Moya-Albor, Jorge Brieva and Hiram Ponce would like to thank the Facultad de Ingeniería of Universidad Panamericana (Campus Mexico City) for all support in this work.

Appendix A

Simulated Annealing in python programming language used for the Traveling Salesman Problem presented in Section 2.2.1.

Listing 1: Traveler Salesman Problem in Python 3

```python
import random
import math
from typing import List, Union, Any

import numpy as np
import matplotlib.pyplot as plt


def choose_new_state(actual_temperature, state, proposal):
    cost_difference = state - proposal
    probability_of_acceptance = abs(math.exp((cost_difference/actual_temperature)))
    if np.random.uniform(1, 0, 1) <= probability_of_acceptance:
        return True, probability_of_acceptance
    else:
        return False, probability_of_acceptance


class Location:
    def __init__(self, x, y):
        self.x = x
        self.y = y


def get_distance(a, b):
    dist = np.sqrt(np.abs(a.x-b.x)**2 + np.abs(a.y - b.y)**2)
    return dist


def get_total_distance(coordinates):
    dist = 0
    for first_city, second_city in zip(coordinates[:-1], coordinates[1:]):
        dist = dist + get_distance(first_city, second_city)
    dist = dist + get_distance(coordinates[0], coordinates[-1])
    return dist


def generate_cities(number_of_cities):
    list_of_cities = []

    for number in range(number_of_cities):
        x = np.random.randint(1, 200)
        y = np.random.randint(1, 200)
        list_of_cities.append(Location(x, y))

    return list_of_cities


def calculate_initial_temperature(num_samples, total_cities):
    deviation = np.array([])
```

Coronary Stenosis Detection using CNN and Simulated Annealing ■ 243

```python
    for sample in range(num_samples):
        deviation = np.append(deviation, [get_total_distance(generate_cities(total_cities))])
    deviation = np.std(deviation)
    return deviation


# Choose the number of total cities
cities = 15
coordinates = []
cities_list = generate_cities(cities)
fig = plt.figure(figsize=(12, 7))
for i in range(len(cities_list)):
    coordinates.append(Location(cities_list[i].x, cities_list[i].y))
    plt.title("Cities")
    plt.scatter(coordinates[i].x, coordinates[i].y)


def graph_solution(cities_coordinates):
    cost_of_solution = get_total_distance(cities_coordinates)
    solution = plt.figure(figsize=(12, 7))
    for actual_city, next_city in zip(cities_coordinates[:-1], cities_coordinates[1:]):
        plt.plot([actual_city.x, next_city.x], [actual_city.y, next_city.y], "b")
    plt.plot([cities_coordinates[0].x, cities_coordinates[-1].x], [cities_coordinates[0].y, cities_coordinates[-1].y], "b")
    for coordinate in cities_coordinates:
        plt.plot(coordinate.x, coordinate.y, "ro")
    plt.title(f"Solution_with_distance_of_{cost_of_solution}")


graph_solution(coordinates)
# Simulated annealing
cost = get_total_distance(coordinates)
for number_of_simulated_annealing in range(3):
    T = calculate_initial_temperature(num_samples=20, total_cities=cities)
    cooling = 0.99
    iterations_per_temperature = int(cities ** 2)
    cost1 = 0
    cost_list = []
    temperatures = []
    while T >= 0.1:
        for i in range(iterations_per_temperature):
            r1, r2 = random.sample(range(len(coordinates)), 2)

            temporal = coordinates[r1]
            coordinates[r1] = coordinates[r2]
            coordinates[r2] = temporal
            # cost
            cost1 = get_total_distance(coordinates)
            accept, prob = choose_new_state(actual_temperature=T, state=cost, proposal=cost1)
            if accept is True:
                cost = cost1
                cost_list.append(cost1)

            else:
                temp = coordinates[r1]
                coordinates[r1] = coordinates[r2]
                coordinates[r2] = temp
```

```
        T = T * cooling

    graph_solution(coordinates)

fig = plt.figure(figsize=(16, 8))
plt.plot(cost_list)
plt.title("Cost_function")
plt.xlabel("Number_of_solution")
plt.ylabel("Cost_function")
plt.show()
```

References

Antczak Karol and Liberadzki Lukasz. (2018). Stenosis detection with deep convolutional neural networks. *MATEC Web Conf.*, 210:04001.

Bas van Stein, Hao Wang and Thomas Bäck. (2019). Automatic configuration of deep neural networks with parallel efficient global optimization. In *2019 International Joint Conference on Neural Networks (IJCNN)*, pp. 1–7.

Basheer Qolomany, Majdi Maabreh, Ala Al-Fuqaha, Ajay Gupta and Driss Benhaddou. (2017). Parameters optimization of deep learning models using particle swarm optimization. In *2017 13th International Wireless Communications and Mobile Computing Conference (IWCMC)*, pp. 1285–1290.

Ben-Ameur. W. (2004). Computing the initial temperature of simulated annealing. *Computational Optimization and Applications*, 29(1): 369–385.

Bin Wang, Bing Xue and Mengjie Zhang. (2020). Particle swarm optimisation for evolving deep neural networks for image classification by evolving and stacking transferable blocks.

Bin Wang, Yanan Sun, Bing Xue and Mengjie Zhang. (2018). Evolving deep convolutional neural networks by variable-length particle swarm optimization for image classification.

Chao Cong, Yoko Kato, Henrique Doria Vasconcellos, Joao Lima and Bharath Venkatesh. (November 2019). Automated stenosis detection and classification in x-ray angiography using deep neural network. In *Proceedings—2019 IEEE International Conference on Bioinformatics and Biomedicine, BIBM 2019*, pp. 1301–1308.

Chi-Hwa Song, Kyunghee Lee and Won Don Lee. (2003). Extended simulated annealing for augmented tsp and multi-salesmen tsp. In *Proceedings of the International Joint Conference on Neural Networks*, volume 3, pp. 2340–2343.

Danilov, V. V., Klyshnikov, K. Y., Gerget, O. M., Kutikhin, A. G. a., Ganyukov, V. I. et al. (2021). Real-time coronary artery stenosis detection based on modern neural networks. *Scientific Reports*, 11(7582).

David P. Williamson and David B. Shmoys. (2010). *The Design of Approximation Algorithms.* Cambridge University Press.

Dowsland, K. A. and Thompson, J. M. (2012). Simulated annealing. *Handbook of Natural Computing*, pp. 1623–1655.

Emmanuel Ovalle-Magallanes, Juan Gabriel Avina-Cervantes, Ivan Cruz-Aceves and Jose Ruiz-Pinales. (2020). Transfer learning for stenosis detection in x-ray coronary angiography. *Mathematics*, 8(9).

Francisco Erivaldo Fernandes Junior and Gary G. Yen. (2019). Particle swarm optimization of deep neural networks architectures for image classification. *Swarm and Evolutionary Computation*, 49: 62–74.

Ivan Cruz-Aceves, Fernando Cervantes-Sanchez and Arturo Hernandez-Aguirre. (2017). *Automatic Detection of Coronary Artery Stenosis Using Bayesian Classification and Gaussian Filters Based on Differential Evolution*, pp. 369–390.

Jong Hak Moon, Da Young Lee, Won Chul Cha, Myung Jin Chung, Kyu-Sung Lee et al. (2021). Automatic stenosis recognition from coronary angiography using convolutional neural networks. *Computer Methods and Programs in Biomedicine*, 198: 105819.

Kirkpatrick, S., Gelatt, C. D. and Vecchi, M. P. (1983). Optimization by simulated annealing. *Science*, 220(4598): 671–680.

Lewis, R. (2007). Metaheuristics can solve sudoku puzzles. *Journal of Heuristics*, 13(4): 387–401.

Lingxi Xie and Alan Yuille. (2017). Genetic cnn.

Masanori Suganuma, Shinichi Shirakawa and Tomoharu Nagao. (2018). A genetic programming approach to designing convolutional neural network architectures. In *Proceedings of the Twenty-Seventh International Joint Conference on Artificial Intelligence, IJCAI-18*, pp. 5369–5373. International Joint Conferences on Artificial Intelligence Organization, 7.

Miguel-Angel Gil-Rios, Igor V. Guryev, Ivan Cruz-Aceves, Juan Gabriel Avina-Cervantes, Martha Alicia Hernandez-Gonzalez et al. (2021). Automatic feature selection for stenosis detection in x-ray coronary angiograms. *Mathematics*, 9(19).

Noraini Mohd Razali. (2015). An efficient genetic algorithm for large scale vehicle routing problem subject to precedence constraints. *Procedia—Social and Behavioral Sciences*, 195: 1922–1931. World Conference on Technology, Innovation and Entrepreneurship.

Seong-Hoon Kim, Zong Woo Geem and Gi-Tae Han. (2020). Hyperparameter optimization method based on harmony search algorithm to improve performance of 1d cnn human respiration pattern recognition system. *Sensors*, 20(13).

Serizawa, T. and Fujita, H. (2020). Optimization of convolutional neural network using the linearly decreasing weight particle swarm optimization.

Talbi, E. -G. (2009). Metaheuristics, from Design to Implementation. John Wiley & Sons, Inc.

Toshihiko Yamasaki, Takuto Honma and Kiyoharu Aizawa. (2017). Efficient optimization of convolutional neural networks using particle swarm optimization. In *2017 IEEE Third International Conference on Multimedia Big Data (BigMM)*, pp. 70–73.

Vishal Passricha and Rajesh Kumar Aggarwal. (2019). Pso-based optimized cnn for hindi asr. *International Journal of Speech Technology*, 22(4): 1123–1133.

Yanan Sun, Bing Xue, Mengjie Zhang and Gary G. Yen. (2019). A particle swarm optimization-based flexible convolutional autoencoder for image classification. *IEEE Transactions on Neural Networks and Learning Systems*, 30(8): 2295–2309.

Yanan Sun, Bing Xue, Mengjie Zhang, Gary G. Yen and Jiancheng Lv. (2020). Automatically designing cnn architectures using the genetic algorithm for image classification. *IEEE Transactions on Cybernetics*, 50(9): 3840–3854.

Yuqiao Liu, Yanan Sun, Bing Xue, Mengjie Zhang, Gary G. Yen et al. (2021). A survey on evolutionary neural architecture search.

Zhichao Lu, Ian Whalen, Yashesh Dhebar, Kalyanmoy Deb, Erik Goodman et al. (2020). Nsga-net: Neural architecture search using multi-objective genetic algorithm (extended abstract). pp. 4750–4754. *In*: Christian Bessiere (ed.). *Proceedings of the Twenty-Ninth International Joint Conference on Artificial Intelligence, IJCAI-20*. International Joint Conferences on Artificial Intelligence Organization, 7 2020. Sister Conferences Best Papers.

Chapter 12

Deep Learning Approach for Detecting COVID-19 from Chest X-ray Images

Murali Krishna Puttagunta,[1] S. Ravi[2,]* and C. Nelson Kennedy Babu[3]

ABSTRACT

Coronavirus disease 2019 (COVID-19) is an infectious disease that begins with flu-like symptoms. COVID-19 began in China and spread rapidly throughout the world. This disease usually results in Pneumonia. Due to the fact that pulmonary infections can be observed via radiography images. This paper focuses on the detection of COVID-19 based on Deep Transfer Learning (DTL) methods by analyzing Chest X-ray (CXR) images. The proposed DTL framework classifies CXR images as COVID-19 infected or normal images. Along with the custom CNN, four different deep Convolutional Neural Networks (CNNs) were used: Vgg-16, ResNet-50, InceptionV3, and

---

[1] Research Scholar, Dept of Computer Science, School of Engineering and Technology, Pondicherry University, India. Email: murali93940@gmail.com

[2] Department of Computer Science, School of Engineering and Technology, Pondicherry University, India.

[3] Computer Science and Engineering, Saveetha School of Engineering, Saveetha Institute of Medical and Technical Sciences, Chennai - 602 105.
Email: cnkbabu63@gmail.com

* Corresponding author: sravicite@gmail.com

MobileNet. The CNN models were trained using CXR datasets collected from open access provided by Kaggle and GitHub. In this study, the classification accuracy of Covid-19 and the normal image is 94%, and the AUC was 0.98. Pre-trained CNN models may be used to support radiologists invalidating their initial screening. This paper studies deep learning approaches for analyzing CXR images to provide health professionals with precise tools for screening COVID-19.

1 Introduction

The COVID-19 is a viral pathogen infection caused by severe acute respiratory syndrome coronavirus 2 (SARS-CoV-2), which causes respiratory diseases in humans (Acter et al., 2020). COVID-19 can be spread through the surface of objects, airborne, in particular, respiratory droplets, and close contact with infected animals (Chan et al., 2020). Coronaviruses come from an extensive family of viruses that cause severe pulmonary illnesses named SARS-CoV2. Several virological and genetic studies have shown that the origins of SARS-CoV2 and MERS-CoV are supposed to be bats. Simultaneously, the intermediate hosts before dissemination to humans are dromedary camels and palm civets, respectively (Hu et al., 2015).

There are two widely used diagnoses. One method is to test for the presence of viral RNA fragments using a nasopharyngeal swab. Following that, the samples are subjected to real-time reverse transcription-polymerase chain reaction (RT-PCR) analysis. A nasal swab or sputum sample may also be used in some circumstances. The results of RT-PCR are typically available within several hours to two days. Another method is the imaging technique. Chest X-ray images (CXR) and computerized tomography (CT) images are commonly used to examine and diagnose COVID-19. Every hospital has an X-ray imaging device, and COVID-19 may be detected using an X-ray image without any specific diagnostic equipment. Physicians used CXR images to diagnose Pneumonia, lung disease, abscesses, and enlarged lymph nodes. Automation of the image acquisition workflow is necessary to prevent the risk of infection while screening the patient. The advantages of imaging techniques are that they may help with screening or expedite diagnosis, mainly when RT-PCR is unavailable (Brunese et al., 2020).

Deep learning (DL) is a subfield of machine learning, specifically artificial intelligence (AI), concerned with algorithms that utilize artificial neural networks (Gianchandani et al., 2020). A variety of medical imaging and deep learning algorithms currently have several medical applications in clinical research. Since deep learning techniques, particularly CNNs, have surpassed humans in various computer vision tasks, Pneumonia and other diseases have already been studied using DL to detect and classify radiography images.

2 Related Work

Ozturk et al., 2020 built the DL network called DarkCovidNet for automatic COVID-19 diagnosis using CXR images. The dataset comprises 500 images of no-finding, 127 images of COVID-19, and 500 images of Pneumonia. Ozturk et al., 2020 used data augmentation to increase the ratio of images per class. The results achieved had a specificity of 0.921, the sensitivity was 0.853, the F1-Score was 0.873, the precision was 0.899, and the accuracy was 0.870.

Oh et al. (Oh et al., 2020) proposed a patch-based deep CNN architecture with a relatively small dataset for COVID-19 diagnosis. Wong and Wang, 2020 developed a deep CNN model named COVID-Net for COVID-19 detection, which achieved a detection accuracy of 92.6%. Luz et al. (Luz et al., 2020) investigated DL architectures for COVID-19 detection. The COVIDx dataset was utilized to illustrate the suggested model's efficiency. The Flat EfficientNet model attained an accuracy of 93.9 per cent and a sensitivity of 96.8 per cent. Using the ResNet50 model, Narin et al., 2020 achieved a testing accuracy of 98 per cent on a dataset of 100 images (50 COVID and 50 normal images). Farooq and Hafeez, 2020 developed a CNN model based on ResNet called COVID-ResNet to categorize COVID-19. However, only 68 COVID-19 CXR images were considered in this study. Hemdan et al., 2020 designed a COVIDX-Net model with X-ray images in consideration. The COVIDX-Net model was trained using seven distinct CNN models and verified using 50 CXR images. Basu and Mitra (Basu et al., 2020) suggested a transfer learning technique for identifying COVID-19-induced abnormalities in CXR images. They employed Gradient Class Activation Maps to recognize the abnormality characteristics in X-ray images and verified their technique using the NIH chest X-ray dataset. This model had an accuracy of 95.3%.

Apostolopoulos and Mpesiana, 2020 trained various pre-trained CNN models on a dataset of 224 COVID-19 images and reached an accuracy of 93.48 per cent for three classes and 98.75 per cent for two classes, respectively. Sethy and Behera (Sethy et al., 2020) classified COVID-19 using a combination of CNN models and a support vector machine (SVM) classifier. According to their analysis, the ResNet50 model with SVM classifier showed the most outstanding performance. Singh et al., 2021 originally classified COVID-19 using CNN variations such as PSO-CNN and GA-CNN. Rajaraman et al. (Rajaraman et al., 2020) used pruning and ensembling to improve the performance of their DL models. The authors devised a method for identifying COVID-19-related CXR anomalies. Additionally, they improved the classification accuracy by doing modality-specific training on pneumonia-related data in addition to pre-training on ImageNet. The best models were then pruned iteratively.

3 CNN Architectures

Recent advances in medical image analysis including detection, segmentation, and classification have been made using DL techniques and CNNs (Puttagunta and Ravi, 2021). DL has demonstrated effectiveness in automating feature-representation learning and gradually eliminating handmade feature engineering. CNN's seek to imitate the human visual cortex system in terms of structure and operation. This multi-layer feature representation allowed CNNs to learn different image features automatically, outperforming handmade feature approaches.

Hubel and Wiesel, 1962 demonstrated that a monkey's visual area might be represented topographically. In the 1980s, the authors (Fukushima, 1980) introduced Neocognitron; a self-organizing neural network considered the precursor of CNN. In LeCun et al. developed contemporary CNN models known as LeNet for handwritten digit recognition. Convolutional layers and the backpropagation method for training became an essential building component of most contemporary CNN architectures. According to Krizhevsky et al., 2012, AlexNet architecture outperformed other techniques in 2012, lowering top-5 errors from 26 to 15.3%. AlexNet is a deep structure version of LeNet trained on 1.2 million images. AlexNet promoted substantial computing resources such as graphics processing units to compensate for the rise in trainable parameters.

Karen Simonyan and Andrew Zisserman of the Oxford Robotics Institute created the Visual Geometry Group Network (VGG) based on the CNN architecture (Simonyan and Zisserman, 2015). The VGGNet performed well on the ImageNet dataset at the ILSVRC2014. To improve the effectiveness of image extraction, the VGGNet used smaller filters of 3 × 3, as opposed to the AlexNet filter of 11 × 11. VGG16 and VGG19 are two variations of this deep network design, each with different layers and depths. Inception network was a 22-layer network that won the 2014 Image net challenge with a top-5 accuracy of 93.3 per cent.

In Inception v3, (Szegedy et al., 2016) several approaches were adopted, including factorized convolutions, smaller convolutions, asymmetric convolutions, auxiliary classifiers, regularization, dimension reduction, and parallelized computation, focus on reducing computing power costs through modifications to prior Inception architectures. ResNet-50 is a ResNet version that features 48 convolutional layers, one MaxPool, and one Average Pool layer (He et al., 2016). Sandler et al. (2018) presented the MobileNetV2 model as a CNN architecture for low-power devices such as smartphones. MobileNets accomplish this critical advantage by minimizing the number of learning parameters and incorporating inverted-residuals-with-linear-bottleneck-blocks to significantly minimize memory usage. Additionally,

MobileNetV2's pre-trained implementation is widely accessible in a variety of popular deep learning frameworks.

4 Proposed Method

In recent years, deep learning methods have demonstrated a steady growth pattern in the medical field. CNN is a deep learning approach used mainly for medical image classification and disease detection. CNN has demonstrated a remarkable performance in visual recognition tasks (Razavian et al., 2014). The CNN's successful classification performance depends on the availability of a vast volume of annotated medical data. The concept of deep transfer learning has been used in different medical applications such as detection of Pneumonia (Chouhan et al., 2014), detection of Tuberculosis (Hooda et al., 2019) and Cervical Histopathology Image Classification (Xue et al., 2020). Figure 1 illustrates the proposed approach for the diagnosis of COVID-19 cases. The model's objective is to identify a given CXR image as COVID-19 or normal. The following sections provide a thorough description of each stage in further depth.

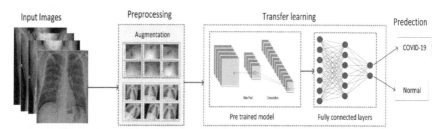

Figure 1: Overview of the COVID-19 diagnostic technique based on chest X-ray images.

4.1 Custom CNN

The custom CNN was comprised of three convolutional blocks, pooling and dense layers. The convolution block comprises a depth-wise separable convolution layer, batch normalization, and a non-linear activation function ReLU (Rajaraman and Antani, 2020). The separable convolution operation is applied to each channel, followed by a point-wise convolution of 1 × 1 kernels.

Convolution of the input signal $x[c, d]$ and the impulse response $h[c, d]$ may be expressed as follows:

$$y[c,d] = x[c,d] * h[c,d] = \sum_{a=-\infty}^{\infty}\sum_{b=-\infty}^{\infty} h[c,d]x[c-a,d-b] \qquad (1)$$

The separable convolution is performing twice of a 1D convolution in the horizontal and vertical directions:

$$y[c,d] = h[c,d] * x[c,d] \qquad (2)$$

If $h[c, d]$ is separable to $(C \times 1)$ and $(1 \times D)$
Then

$$y[c,d] = (h_1[c].h_2[d]) * x[c,d] = h_2[d] * (h_1[c] * x[c,d]) \\ = h_1[c] * (h_2[d] * x[c,d]) \qquad (3)$$

where * indicates convolution.

In CNN, the following mathematical expression represents the convolution operation between the image $I(x, y)$ and the filter $K(x, y)$.

$$conv(I,K)_{xy} = \Sigma_{i=1}^{n_H} \Sigma_{j=1}^{n_w} \Sigma_{m=1}^{n_c} K_{i,j,m} * I_{x+i-1, y+i-1, m} \qquad (4)$$

where n_H is the height of the image, n_w is the width of the image n_c is the number of channels Depthwise Separable convolution mathematically expressed as [32]:

$$pointwiseconv(I,K)_{xy} = \Sigma_m^{n_c} K_m * I_{(x,y,m)} \qquad (5)$$

$$depthwiseconv(I,K)_{xy} = \Sigma_{i=1}^{n_H} \Sigma_{j=1}^{n_w} K_{i,j} \odot I_{x+i, y+j} \qquad (6)$$

$$SepConv(I_H, I_v, k)_{x,y} = pointwiseconv_{xy}(I_H, depthwiseConv_{xy}(I_v, k)) \qquad (7)$$

Assume that the size of image is $S \times S$, the size of the convolutional kernel is $f \times f \times N_c$, and N_{oc} is the number of output channels [33]

The computational cost and parameters of the traditional convolution are:

$$computation cost = S \times S \times f \times f \times N_c \times N_{oc} \qquad (8)$$

$$parameters = f \times f \times N_c \times N_{oc} \qquad (9)$$

The computational cost and parameters of depthwise separable convolution are:

$$computational cost = S \times S \times f \times f \times N_c + N_c \times N_{oc} \times S \times S \qquad (10)$$

$$parameters = S \times S \times N_c + N_c \times N_{oc} \qquad (11)$$

where S is the width and height of the input image, f is the filter width and height of filter, N_c is the number of input channels and N_{oc} is the number of output channels.

Figure 2 shows the internal operations of a depthwise separable CNN. Separable convolution leads to fewer model parameters and lesser overfitting than the traditional convolution (Zhang et al., 2019; Wang et al., 2018). Depthwise separable convolutions are used in Xception and MobilNets to

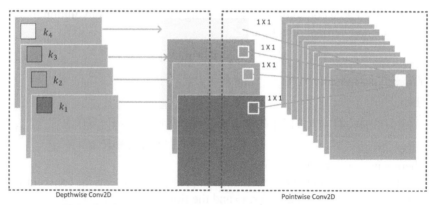

Figure 2: The detailed internal operation of a Depthwise Separable CNN.

build image classification models that outperform similar architectures. The custom CNN architectural framework is in Fig. 3(a). Zero padding is used in the convolution layers to ensure that the middle layers' feature map dimensions equal the initial input size. The separable convolution layer filter size is 5 × 5. The max-pooling layer comes after every convolutional block. The custom CNN model was added with a pooling layer, the global average pooling layer (GAP), followed by dropout. Finally, a dense layer was added with Softmax activation for output prediction.

4.2 Deep Transfer Learning

Deep transfer learning (DTL) is reusing the network layer weights of pre-trained CNN models for a new domain problem. Transfer learning techniques were presented due to COVID-19's limited dataset collection. CNN's are trained on many annotated image datasets (such as ImageNet) and then transferred and reprocessed for an unfamiliar task (He et al., 2016). Deep transfer learning is an effective technique that has demonstrated substantial results in medical applications such as COVID-19 detection (Sandler et al., 2018), pneumonia detection (Razavia et al., 2014), lung nodule detection (Chouhan et al., 2014), and malaria parasite detection (Hooda et al., 2019). In this study, Four distinct CNN models were considered for the classification of normal and COVID-19 affected CXR images that are VGG-16, (Simonyan and Zisserman, 2015) ResNet-50, (He et al., 2016) Inception-V3, (Szegedy et al., 2016) MobilNet (Howard et al., 2017; Sandler et al., 2018).

A huge ImageNet database was used to pre-train the networks before progressive training on CXR images. The pre-trained CNN models with their ImageNet weights (https://github.com/fchollet/deep-learning-models/blob/master/imagenet_utils.py,2020) are considered and truncated at their fully

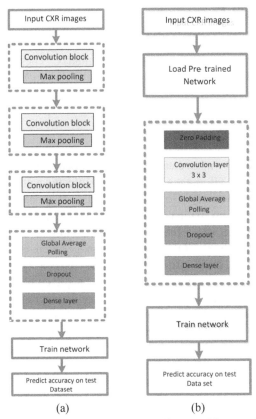

Figure 3: (a) custom CNN model Architecture (b) Architecture of the pre-trained CNN model.

connected layers. The following layers replaced the truncated final layers: (i) zero-padding, (ii) convolutional layers with 3 × 3 filters and 1024 feature maps, (iii) Global Average Pooling layer (GAP), (iv) dropout layer, and (v) final dense layer with a softmax activation function (Rajaraman et al., 2020). A brief mechanism for deep transfer learning is shown in Fig. 3(b).

Models are fine-tuned with weight improvements by the SGD optimization algorithm for minimizing the categorical cross-entropic loss to the CXR classification. The hyperparameters of the pre-trained CNN models include L2-weight decay, learning rate and momentum (Bergstra and Bengio, 2012). Table 1 shows the values of the hyperparameters.

Table 1: Hyper parameter values.

Model	Learning rate	Momentum	L2 decay
Pretrained CNNs	1e-3	0.95	1e-6
Custom CNN	1e-3	0.90	1e-5

5 Results and Discussion

The evaluation of models in deep transfer learning was done in terms of the performance metrics of the confusion matrix. These metrics include (i) accuracy (ii) Recall (iii) Area under the ROC curve (AUC) (iv) Precision and (v) F1-Score.

5.1 Dataset

This paper uses CXR image datasets from two open sources for the diagnosis of COVID-19. The CXR images of COVID-19 infected patients were collected from the GitHub repository (Cohen et al., 2003). This dataset comprises 382 images collected from patients infected with COVID-19, Pneumocystis, Streptococcus, SARS and other types of Pneumonia. In this dataset, the 142 posteroanterior view COVID-19 CXR images were considered. Normal CXR images were taken from the Kaggle dataset (RSNA Pneumonia Detection Challenge|Kaggle." https://www.kaggle.com/c/rsna-pneumonia-detection-challenge/overview, 2020). The Kaggle dataset consists of 8851 normal images and 6012 pneumonia images. Dataset-1 was created by merging 142 COVID-19 images from Github and 142 regular CXR images from Kaggle to form a balanced data set. The Dataset-1 was split into 224 images (80% total) for training and 60 images (20% of total) for testing the models. In the training data, 22 randomly allocated images were used for validation.

The available COVID-19 CXR image dataset has fewer images (only 284), which results in overfitting when the neural network was trained with this tiny dataset. Data augmentation methods are applied to increase the size of the original dataset. With the lack of a benchmark dataset for COVID-19 CXR images, the augmentation approach was used. The augmentation techniques are histogram equalization, horizontal flipping, rotating images in three different angles, and translation so that each image can generate six images. The augmented COVID-19 CXR images contain $142 \times 7 = 994$ images (six augmented images and one original image).

Mendeley has released a dataset of augmented COVID-19 CXR images (Alqudah et al., 2020). This dataset consists of Augmented COVID-19 CXR images. Based on the augmented data set concept, we created a new dataset. Dataset-2 consists of augmented COVID-19 images and augmented normal images. Figure 4 illustrates the original and augmented CXR Images from Dataset-1 and Dataset-2. The dataset details are enlisted in Table 2.

Deep Learning Approach for Detecting COVID-19 from Chest X-ray Images ■ 257

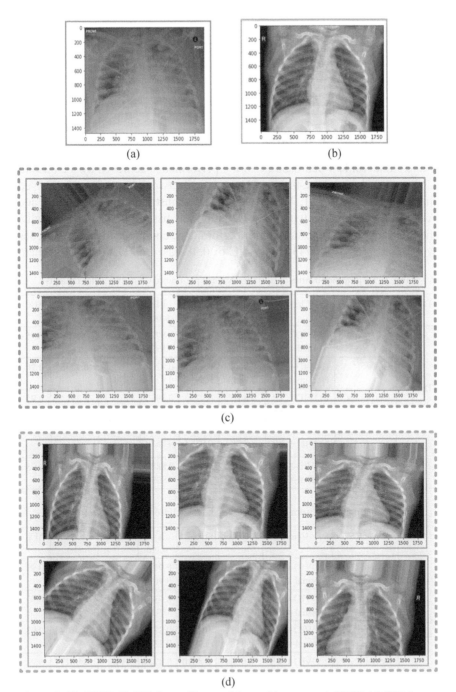

Figure 4: (a) COVID-19 CXR image (b) normal image (c) augmented COVID-19 CXR images (d) augmented normal images. These images are from Dataset-1 and Dataset-2.

Table 2: Data set.

	Class	Training	Test	Total
Dataset-1	Covid-19	112	30	284
	Normal	112	30	
Dataset-2	Covid-19	994	142	2272
	Normal	994	142	

5.2 Experimental Results

Table 3 and Table 4 demonstrate of the performance achieved by the custom CNN and DTL based pre-trained CNNs using the two datasets. It was observed that based on Dataset-1. VGG-16 model demonstrated good accuracy and AUC values compared with other CNN models based on Dataset 1. Finally, the ResNet50 model shows superior performance compared to all other modes with an accuracy of 0.9435 and AUC of 0.98 based on Dataset 2. Excellent

Table 3: Performance metrics of Custom CNN and pre-trained CNN models for Dataset-1.

Dataset	Model	Class	precision	recall	F1-score	Support	Accuracy (%)	AUC
Data Set-1	Custem CNN	COVID-CXR	0.9600	0.8000	0.8727	30	0.8829	0.95
		Normal–CXR	0.8286	0.9667	0.8923	30		
	Vgg16	COVID-CXR	0.9062	0.9467	0.9265	30	**0.9233**	**0.96**
		Normal–CXR	0.9543	0.8918	0.9219	30		
	InceptionV3	COVID-CXR	0.8677	0.9200	0.8930	30	0.8913	0.95
		Normal–CXR	0.9157	0.8567	0.8852	30		
	MobileNet	COVID-CXR	0.9259	0.8333	0.8772	30	0.8833	0.95
		Normal–CXR	0.8485	0.9333	0.8889	30		
	ResNet-50	COVID-CXR	1.000	0.800	0.8889	30	0.8934	0.95
		Normal–CXR	0.833	1.000	0.9091	30		

Table 4: Performance metrics of Custom CNN and pre-trained CNN models for Dataset-2.

Dataset	Model	Class	precision	recall	F1-score	Support	Accuracy (%)	AUC
Data set-2	Custom CNN	COVID-CXR	0.8871	0.9392	0.9124	142	0.8968	0.95
		Normal–CXR	0.9212	0.8943	0.9075	142		
	Vgg16	COVID-CXR	0.9361	0.9295	0.9327	142	0.9330	0.97
		Normal–CXR	0.9300	0.9366	0.933	142		
	InceptionV3	COVID-CXR	0.9131	0.8449	0.8776	142	0.9164	0.93
		Normal–CXR	0.8589	0.9056	0.8816	142		
	MobileNet	COVID-CXR	0.9604	0.8010	0.8734	142	0.9023	0.93
		Normal–CXR	0.8287	0.9665	0.8923	142		
	ResNet-50	COVID-CXR	0.9562	0.9295	0.9426	142	**0.9435**	**0.98**
		Normal–CXR	0.9315	0.9577	0.9444	142		

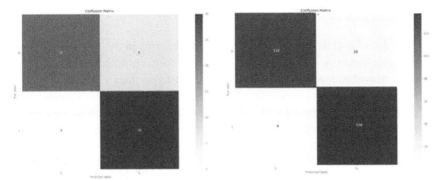

Figure 5: Confusion matrix of VGG-16 model (Dataset-1) and ResNet-50 (Dataset-2).

VGG16

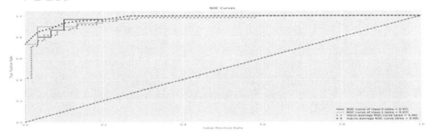

ResNet-50

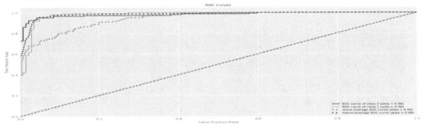

Figure 6: AUC curves of VGG-16, and ResNet-50.

performance models Confusion matrix and ROC curves are shown in Fig. 5 and Fig. 6.

Table 5 found that proposed pre-trained CNN models outperformed the Catack (Catak, 2020) transfer learning models.

Table 5: The proposed pre-trained CNN model's performance was compared to that of other transfer learning models.

	VGG-16	ResNet	Inception V3
Catack, 2020	0.8	0.5	0.6
Proposed Pretrained CNN models	0.933	0.902	0.89

6 Visualization of CNN Layers

CNN's detect and learn edge and colour-like features in the first convolution layer. The network is learning to identify more complicated features in deeper convolutional layers. Later layers construct their characteristics by combining features from earlier layers (Rahman et al., 2020). Visualization of a given input of images with intermediate activations displays the feature maps produced by different convolution and pooling layers and a network view by different filters. Figure 7 shows the feature map of VGG-16 block1_conv1with 64 filters (a), block2_conv2 with 128 filters (b), block3_conv3 with 256 filters (c), and block3_conv4 with 512 filters (d).

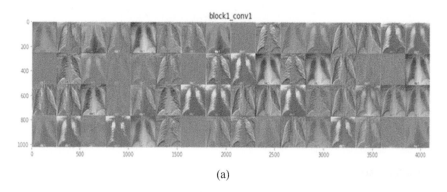

(a)

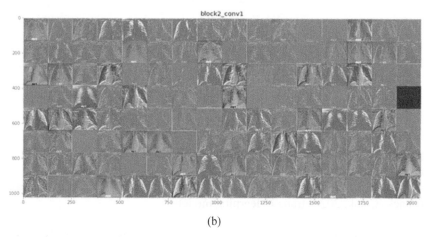

(b)

Fig. 7 contd. ...

...*Fig. 7 contd.*

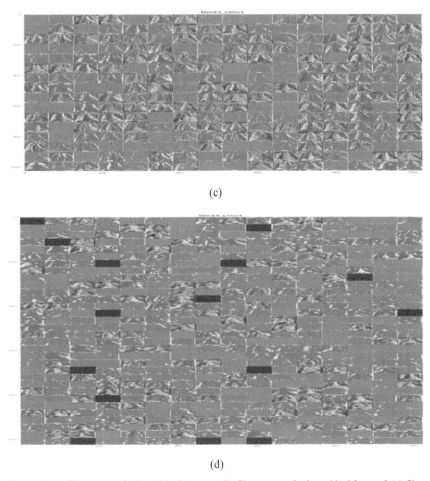

Figure 7: (a) filter pattern for layer block1_conv1 (b) filter pattern for layer block2_conv2 (c) filter pattern for layer block3_conv1 (d) filter pattern for layer block4_conv1.

7 Conclusion

This research focused on developing deep learning techniques that healthcare practitioners or radiologists may utilize to provide a precise and easy diagnosis of COVID-19 patients. To detect and categorize COVID-19 cases from CXR pictures, we proposed a deep transfer learning model. Due to the developing nature of COVID-19 as a worldwide pandemic, it is challenging to collect vast data sets for systematically training deep neural networks. The technique of augmentation was used to increase the number of images in the data collection. After augmentation, the covid19 CXR images improved from 142 to 994. The suggested approach makes use of a limited dataset of images. As the dataset images number increases, the suggested model may provide more dependable accuracy. Generative Adversarial Networks (GAN) may be used to generate additional pictures. A DTL system based on deep learning dramatically improves the detection of the COVID-19 disease using CXR images.

References

Acter, T., Uddin, N., Das, J., Akhter, A., Choudhury, T. R., Kim, S. et al. (2020). Evolution of severe acute respiratory syndrome coronavirus 2 (SARS-CoV-2) as coronavirus disease 2019 (COVID-19) pandemic: A global health emergency. *Sci. Total Environ.*, 730: 138996, doi: 10.1016/j.scitotenv.2020.138996.

Alqudah, S., Ali Mohammad, Qazan. (2020). Augmented COVID-19 X-ray Images Dataset. vol. 4. Mendeley Data, Mar. 26, 2020, doi: 10.17632/2FXZ4PX6D8.4.

Apostolopoulos, I. D. and Mpesiana, T. A. (2020). Covid-19: Automatic detection from X-ray images utilizing transfer learning with convolutional neural networks. *Phys. Eng. Sci. Med.*, 0123456789: 1–6, doi: 10.1007/s13246-020-00865-4.

Basu, S., Mitra, S. and Saha, N. (2020). Deep Learning for Screening COVID-19 using Chest X-Ray Images, pp. 1–6 [Online]. Available: http://arxiv.org/abs/2004.10507.

Bergstra, J. and Bengio, Y. (2012). Random search for hyper-parameter optimization. *J. Mach. Learn. Res.*, 13: 281–305.

Brain, G., Brain, G. and Gomez, A. N. (2018). Depthwise Separable Convolutions for. *Iclr*, pp. 1–10.

Brunese, L., Mercaldo, F., Reginelli, A. and Santone, A. (2020). Explainable deep learning for pulmonary disease and coronavirus COVID-19 detection from X-rays. *Comput. Methods Programs Biomed.*, 196, doi: 10.1016/j.cmpb.2020.105608.

Catak, F. O. (2020). Transfer Learning Based Convolutional Neural Network For COVID-19 Detection With X-Ray Images. June, 2020.

Chan, J. F. W. et al. (2020). A familial cluster of pneumonia associated with the 2019 novel coronavirus indicating person-to-person transmission: A study of a family cluster. *Lancet*, 395(10223): 514–523, doi: 10.1016/S0140-6736(20)30154-9.

Chouhan, V. et al. (2020). A novel transfer learning based approach for pneumonia detection in chest X-ray images. *Appl. Sci.*, 10(2), doi: 10.3390/app10020559.

Cohen, J. P., Morrison, P. and Dao, L. (2020). COVID-19 Image Data Collection. Mar. 2020, Accessed: Jun. 28, 2020. [Online]. Available: http://arxiv.org/abs/2003.11597.

Deep-learning-models/imagenet_utils.py at master. fchollet/deep-learning-models. https://github.com/fchollet/deep-learning-models/blob/master/imagenet_utils.py (accessed Jun. 28, 2020).

Depthwise separable convolutions for machine learning - Eli Bendersky's website. https://eli.thegreenplace.net/2018/depthwise-separable-convolutions-for-machine-learning/#id1 (accessed Jul. 08, 2020).

Farooq, M. and Hafeez, A. (2020). COVID-ResNet: A Deep Learning Framework for Screening of COVID19 from Radiographs [Online]. Available: http://arxiv.org/abs/2003.14395.

Fukushima, K. (1980). Neocognitron: A self-organizing neural network model for a mechanism of pattern recognition unaffected by shift in position. *Biol. Cybern.*, 36(4): 193–202, doi: 10.1007/BF00344251.

Gianchandani, N., Jaiswal, A., Singh, D., Kumar, V., Kaur, M. et al. (2020). Rapid COVID-19 diagnosis using ensemble deep transfer learning models from chest radiographic images. *J. Ambient Intell. Humaniz. Comput.*, 0123456789, doi: 10.1007/s12652-020-02669-6.

He, K., Zhang, X., Ren, S. and Sun, J. (2016). Deep residual learning for image recognition. *Proc. IEEE Comput. Soc. Conf. Comput. Vis. Pattern Recognit.*, 2016-Decem: 770–778, doi: 10.1109/CVPR.2016.90.

Hemdan, E. E. -D., Shouman, M. A. and Karar, M. E. (2020). COVIDX-Net: A framework of deep learning classifiers to diagnose COVID-19 in X-Ray images. *arXiv* [Online]. Available: http://arxiv.org/abs/2003.11055.

Hooda, R., Mittal, A. and Sofat, S. (2019). Automated TB classification using ensemble of deep architectures. *Multimed. Tools Appl.*, 78(22): 31515–31532, doi: 10.1007/s11042-019-07984-5.

Howard, A. G. et al. (2017). MobileNets: Efficient Convolutional Neural Networks for Mobile Vision Applications [Online]. Available: http://arxiv.org/abs/1704.04861.

Hu, B., Ge, X., Wang, L. F. and Shi, Z. (2015). Bat origin of human coronaviruses Coronaviruses: Emerging and re-emerging pathogens in humans and animals Susanna Lau Positive-strand RNA viruses. *Virol. J.*, 12(1): 1–10, doi: 10.1186/s12985-015-0422-1.

Hubel, D. H. and Wiesel, T. N. (1962). Receptive fields, binocular interaction and functional architecture in the cat's visual cortex. *J. Physiol.*, 160(1): 106–154, doi: 10.1113/jphysiol.1962.sp006837.

Krizhevsky, A., Sutskever, I. and Hinton, G. E. (2012). ImageNet classification with deep convolutional neural networks. In *the 25th International Conference on Neural Information Processing Systems*, pp. 1097–1105, doi: 10.1145/3065386.

LeCun, Y. et al. (1989). Backpropagation applied to digit recognition. *Neural Computation*, 1(4): 541–551.

Luz, E., Silva, P. L., Silva, R., Silva, L., Moreira, G. et al. (2020). Towards an effective and efficient deep learning model for COVID-19 patterns detection in X-ray images, pp. 1–10, Apr. 2020, Accessed: Aug. 01, 2020. [Online]. Available: http://arxiv.org/abs/2004.05717.

Narin, A., Kaya, C., Pamuk, Z., Ali Narin, Ceren Kaya et al. (2020). Automatic detection of coronavirus disease (COVID-19) using X-ray images and deep convolutional neural networks. *arXiv Prepr. arXiv2003.10849.* [Online]. Available: https://arxiv.org/abs/2003.10849.

Oh, Y., Park, S. and Ye, J. C. (2020). Deep learning COVID-19 features on CXR using limited training data sets. *IEEE Trans. Med. Imaging*, 0062(c): 1–1, doi: 10.1109/tmi.2020.2993291.

Ozturk, T., Talo, M., Yildirim, E. A., Baloglu, U. B., Yildirim, O. et al. (2020). Automated detection of COVID-19 cases using deep neural networks with X-ray images. *Comput. Biol. Med.*, 121(April): 103792, doi: 10.1016/j.compbiomed.2020.103792.

Puttagunta, M. and Ravi, S. (2021). Medical image analysis based on deep learning approach. *Multimed. Tools Appl.*, 80(16): 24365–24398, doi: 10.1007/s11042-021-10707-4.

Rahman, T., Chowdhury, M. E. H. and Khandakar, A. (2020). Transfer learning with deep convolutional neural network (CNN) for pneumonia detection using chest X-ray. *Appl. Sci.*, 10(9): 1–19, 3233; https://doi.org/10.3390/app10093233.

Rajaraman, S. and Antani, S. K. (2020). Modality-specific deep learning model ensembles toward improving TB detection in chest radiographs. *IEEE Access*, 8: 27318–27326, doi: 10.1109/ACCESS.2020.2971257.

Rajaraman, S., Kim, I. and Antani, S. K. (2020). Detection and visualization of abnormality in chest radiographs using modality-specific convolutional neural network ensembles. *PeerJ*, 2020(3), doi: 10.7717/peerj.8693.

Rajaraman, S., Siegelman, J., Alderson, P. O., Folio, L. S. L. R., Folio, L. S. L. R. et al. (2020). Iteratively pruned deep learning ensembles for COVID-19 detection in chest X-rays. *IEEE Access*, 8: 1–10, doi: 10.1109/ACCESS.2020.3003810.

Razavian, A. S., Azizpour, H., Sullivan, J. and Carlsson, S. (2014). CNN features off-the-shelf: An astounding baseline for recognition. *IEEE Comput. Soc. Conf. Comput. Vis. Pattern Recognit. Work.*, pp. 512–519, doi: 10.1109/CVPRW.2014.131.

RSNA Pneumonia Detection Challenge | Kaggle. https://www.kaggle.com/c/rsna-pneumonia-detection-challenge/overview (accessed Jun. 28, 2020).

Sandler, M., Howard, A., Zhu, M., Zhmoginov, A., Chen, L. C. et al. (2018). MobileNetV2: Inverted residuals and linear bottlenecks. *Proc. IEEE Comput. Soc. Conf. Comput. Vis. Pattern Recognit.*, pp. 4510–4520, doi: 10.1109/CVPR.2018.00474.

Sethy, P. K., Behera, S. K., Ratha, P. K. and Biswas, P. (2020). Detection of coronavirus disease (COVID-19) based on deep features and support vector machine. *Int. J. Math. Eng. Manag. Sci.*, 5(4): 643–651, doi: 10.33889/IJMEMS.2020.5.4.052.

Simonyan, K. and Zisserman, A. (2015). Very deep convolutional networks for large-scale image recognition. In *3rd International Conference on Learning Representations, ICLR 2015*, pp. 1–14.

Singh, D., Kumar, V. and Kaur, M. (2021). Densely connected convolutional networks-based COVID-19 screening model. *Appl. Intell.*, 51(5): 3044–3051, doi: 10.1007/s10489-020-02149-6.

Szegedy, C., Vanhoucke, V., Ioffe, S., Shlens, J., Wojna, Z. et al. (2016). Rethinking the inception architecture for computer vision. In *the IEEE Computer Society Conference on Computer Vision and Pattern Recognition*, 2016-Decem., pp. 2818–2826, doi: 10.1109/CVPR.2016.308.

Wang, G., Yuan, G., Li, T. and Lv, M. (2018). An multi-scale learning network with depthwise separable convolutions. *IPSJ Trans. Comput. Vis. Appl.*, 10(1): 1–8, doi: 10.1186/s41074-018-0047-6.

Wang, L. and Wong, A. (2020). COVID-Net: A Tailored Deep Convolutional Neural Network Design for Detection of COVID-19 Cases from Chest X-Ray Images, pp. 1–12 [Online]. Available: http://arxiv.org/abs/2003.09871.

Xue, D. et al. (2020). An application of transfer learning and ensemble learning techniques for cervical histopathology image classification. *IEEE Access*, 8: 1–1, doi: 10.1109/access.2020.2999816.

Zhang, T., Zhang, X., Shi, J. and Wei, S. (2019). Depthwise separable convolution neural network for high-speed SAR ship detection. *Remote Sens.*, 11(21), doi: 10.3390/rs11212483.

Index

A

ADHD 34–39, 41–43, 47, 48
Affective 133, 136–138, 144, 147, 150
Arrhythmia detection 121, 127
Artificial neural network 156, 158, 159, 162, 167
Automatic neural architecture search 229, 230, 236, 237, 241

B

Behavioral 133, 136, 137, 139, 141, 144, 147, 150

C

CADx 197, 198, 204
Cardiac arrhythmia 121
Cardiac MRI 1–3, 25
Classification 34–37, 39, 40, 42, 43, 47–49, 249–252, 254, 255
Cognitive 133, 135–139, 144, 147, 150
Convolution neural networks 11
Convolutional neural network 92, 97, 99, 155–158, 167, 227–232, 236, 238, 241
Covid-19 155–158, 162, 163, 167, 248–250, 252, 254, 256–258, 261, 262

D

Data Augmentation 35–39
Deep learning 2, 9, 10, 12, 34–37, 39, 47, 48, 52–54, 68, 69, 75–77, 83, 87, 133, 137, 142, 146, 147, 150, 170, 171, 176, 177, 185, 187, 190, 196–200, 202–210, 215, 219–221, 248, 249, 252, 254, 261, 262
Deep learning in healthcare 127

E

Ensemble 52–57, 59, 60, 62–69

F

fmri classification 133–135, 137, 142, 144, 147, 149, 150

G

GRU 134, 136, 137, 144–151

H

Human body 76, 77

L

LSTM 134, 136, 137, 144–151
Lung CT-Scan 155

M

Medical image 75–77, 81–87
Medical imaging 2, 7, 9, 15, 24, 25, 28, 109–111, 113, 115–117, 125, 127
Microscopic image 196, 197, 199–202, 204, 206, 207, 210, 211, 213, 216, 217, 219–221

N

Network training methods 19
Neural networks 109–120

P

Panoramic radiographs 172, 174, 178–181, 184–186, 189

R

Reconstructed image features 155, 156, 158, 159, 166
Red lesion 91, 92, 97–99, 101, 103, 104
Resting 133, 136–140, 142–144, 150
rs-fMRI 136, 138, 142–144, 146, 147

S

Segmentation 1–8, 10–20, 23–28, 52–63, 67–69
Short-axis cine 1–3, 5, 15, 21, 28
Simulated annealing 227–230, 233–239, 241

T

Task classification 134, 144, 146, 148, 149
Task-based fmri 138, 150
Tooth segmentation and numbering 184, 187
Training data 76, 78, 86, 87
Transfer learning 248, 250, 252, 254–256, 259, 261

U

Ulcer 91, 92, 94–96, 99, 101, 103
U-nets 7–11, 16–18, 21–23, 28

W

Wireless capsule endoscopy 91, 92

Y

YOLOV4 215–219, 221